A NEW COURSE OF
PLANTS AND ANIMALS

BOOK I

A NEW COURSE OF
PLANTS AND ANIMALS

BY

M. A. GRIGG, B.Sc.

Senior Science Mistress, Ealing Grammar School for Girls
Formerly Lecturer in Biology, Dudley Training College for Teachers

BOOK I

CAMBRIDGE
AT THE UNIVERSITY PRESS
1959

CAMBRIDGE
UNIVERSITY PRESS

University Printing House, Cambridge CB2 8BS, United Kingdom

Published in the United States of America by Cambridge University Press, New York

Cambridge University Press is part of the University of Cambridge.

It furthers the University's mission by disseminating knowledge in the pursuit of
education, learning and research at the highest international levels of excellence.

www.cambridge.org
Information on this title: www.cambridge.org/9781107650701

© Cambridge University Press 1956

First published 1956
First edition 1956
Reprinted 1959
First paperback edition 2014

A catalogue record for this publication is available from the British Library

ISBN 978-1-107-65070-1 Paperback

PREFACE

BIOLOGY is essentially a study of life, and not a book study. The best lessons are those which can be taken out-of-doors where living things can be studied in their natural surroundings. Unfortunately, very few of us can do this during school hours, so we must bring as many specimens as possible to school for examination, and then study them in their natural surroundings when we can.

In Secondary Grammar Schools, the Biology syllabus is sometimes ruled by the syllabus of the G.C.E. examinations, and general Nature Study is often neglected. This book is intended to guide the studies of boys and girls in Nature Study during the first three years of their Grammar School life, so that they may have a sound general knowledge of the subject before beginning their G.C.E. work.

The first course of *Plants and Animals* was a re-issue of the Biology chapters from *Elementary Science*, which was written primarily for Secondary Modern Schools. The scientific terms were omitted so that all could read and understand the subject. This new course, based on the Biology chapters of *Modern Science*, has been entirely re-written, and some scientific terms have been introduced which will be useful for pupils who intend to take the G.C.E. examination. Where scientific terms have been used, simple descriptions have also been given so that the book is suitable for all boys and girls in Secondary Modern Schools.

With the help of Book I you can study plants and animals which are found in habitats familiar to most boys and girls.

5

Pond life can be studied anywhere, as the plants and animals live quite well together in balanced aquaria.

Many pupils will have access to gardens or parks where many common animals, birds and trees may be found. The flowers chosen for study are found almost anywhere, and most boys and girls are familiar with the animals described in chapter 5.

As far as possible, this book should be read with the living specimens at hand. Reading should guide observation. If animals are to be studied which cannot be brought to school, the teacher should illustrate the lesson by showing pictures (with or without an epidiascope), film strips or films. Visits to a zoological garden or a museum would make the work more interesting.

M. A. GRIGG

28 April 1955

ILLUSTRATIONS

The following were drawn by Miss J. B. S. Willmore: 20 (*c*), 23, 26 (*a* and *b*), 30, 48 (*d*), 96 (*b*), 98 (*c* and *d*), 100, 101, 103–6, 108–10, 113–15, 118, 120–3, 125. The rest are the work of the late Mr J. C. Hill.

CONTENTS

7

CHAPTER 1

THE AQUARIUM

You cannot study living plants and animals unless you watch them closely. Most of us, unfortunately, have no time to study living things in their natural surroundings. We may frequently go out on nature study excursions, but to study animals and plants properly, we must watch them day by day. As we cannot go out every day to make our observations, we must bring the living things into the classroom, where we can watch them.

Pond life is a very fascinating study in school, as the animals and plants can be kept under natural conditions. If we are to study the animals and plants that live in a pond, we must first of all set up an AQUARIUM in which we shall keep them, and then we can go on a fishing expedition to catch as many animals as we can find.

Setting up an aquarium

You may already have a large aquarium in your classroom, but if this is not so, you may use old glass accumulator tanks or two-pound jam jars. Before setting up the aquarium, visit a pond, and collect various kinds of pond weed. Collect some small stones and shingle, wash them thoroughly, and place them on the bottom of your aquarium. Small stones give hiding places for small animals, and also allow the decaying animal and plant matter to fall between them. This decaying matter is easily washed away from the stones when you are cleaning out the aquarium. Sand, on the other hand, holds all the decaying matter, and gives off a very unpleasant smell when the aquarium is cleaned out. Place one or two large stones on top of the smaller ones, and then fill up your aquarium with water. Tap water may be used, provided that

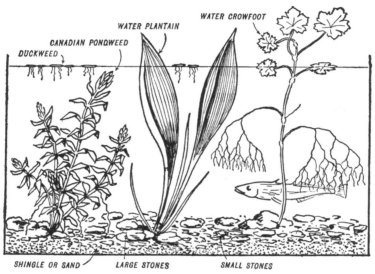

Fig. 1. *An aquarium*

it does not contain too much chlorine. For certain animals (see Chapter 2), you must put some pond water into your aquarium.

During the daytime the green parts of plants give out oxygen, which dissolves in the water, and is breathed in by many animals that live in the water. In return, the animals breathe out carbon dioxide, which the plants use up when making their food (see Book II). We say that the plants

12

AERATE the aquarium. It is obvious therefore that we should put some water weeds into our aquarium. Most of the submerged water plants have only a few or no roots. Place these weeds in your aquarium, with one end of the weed held down by one of the large stones. Plants with underground stems and roots must be planted in a layer of shingle (Fig. 1).

If animals which you have caught in running water, or large animals, are kept in an aquarium, you may have to aerate the water artificially, or keep a constant supply of running water into and out of the aquarium. The only satisfactory method of artificially aerating an aquarium is to use an electric aerator.

Keeping an aquarium clean

If a proper balance between the animals and plants is reached, it should not be necessary to clean out an aquarium too frequently. You may keep an aquarium going for several years, if you look after it carefully. You must frequently wipe the inside of the glass with a clean piece of rag or scrape it with a razor blade, to get rid of the green film which grows there. Do not expect the snails to do this for you. Snails may eat a little of this green stuff, or ALGA, as it is called, but they prefer eating the leaves of the pond weeds. The water which has evaporated must be replaced by siphoning (see Appendix B) water carefully into the tank without stirring up the decaying matter from the bottom. If, in your aquarium, you keep animals such as pond snails and caddis-fly larvae which eat weeds, you must add fresh water-weed to replace that which has been eaten.

Some common pond weeds

When you visit a pond, stream or canal to collect your water plants, you will notice that some plants grow completely under the water. They are TOTALLY SUBMERGED, whilst other plants grow partly in and partly out of the water. They are PARTIALLY SUBMERGED. In addition to these plants you may find some very small plants, called duckweed, floating on top of the water. Collect and keep as many different kinds of water plants as you can find. The following descriptions should help you to identify them.

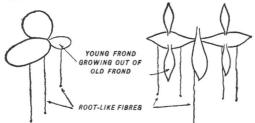

YOUNG FROND
GROWING OUT OF
OLD FROND

ROOT-LIKE FIBRES

Fig. 2. *Round-leaved duckweek and ivy-leaved duckweed*

Floating plants

Duckweed. There are several kinds of duckweed, but all of them are floating plants without distinct stems or real leaves. The plant is made up of leaf-like fronds, which may be separate, or two or three may stick together by their edges. If you look closely at the fronds you will see a fine root-like fibre, about 1 inch long, growing from the frond, downwards into the water. New fronds grow out of the edges of the old fronds. In ivy-leaved duckweed (Fig. 2) two young fronds grow in opposite directions out of the old frond. Tiny flowers sometimes grow out of the fronds. A little duckweed floating on the top of your aquarium is useful, as it shades the surface, and gives a hiding place for small animals.

14

Submerged plants

The flowers of totally submerged plants are usually very small, and are not often seen. These plants usually multiply rapidly, by means of branches, which break off from the parent plant, and continue to grow. If you look closely at some of these plants during the autumn, you will see buds, with leaves packed closely together, growing out of the stems. These WINTER BUDS (see Fig. 3), as they are called, break away from the parent plant, fall to the bottom of the pond, and remain there until the following spring, when they will grow into new plants.

Canadian or American pondweed. This plant is perhaps the most useful one to keep in an aquarium as it multiplies rapidly and freely gives out oxygen. The stems may be very long and branched, and out of them white thread-like roots may grow to anchor the plant in the mud or shingle. These roots which grow out of stems are called ADVENTITIOUS roots. The leaves are arranged in whorls of three (Fig. 3a). Place this plant in bunches with one end of the bunch under a large stone. In the autumn, winter buds are formed.

Water milfoil. Milfoil is a common plant in still water. Its finely divided leaves, which grow out of the stem in whorls of four, make it a favourite plant in an aquarium, because it looks well (Fig. 3c). It also grows quickly and readily gives off oxygen. You may be fortunate enough to see the spike of small, greenish flowers of the spiked milfoil, which grows a few inches out of the water. Place one end of a piece of this plant under a stone in your aquarium and it will soon grow roots. This plant has a root which creeps through the mud at the bottom of a pond. In the autumn, winter buds are formed.

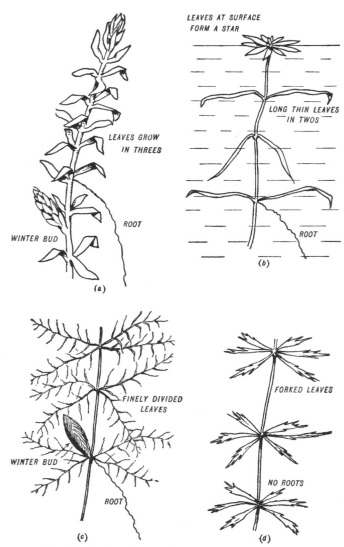

LEAVES AT SURFACE
FORM A STAR

LEAVES GROW
IN THREES

LONG THIN LEAVES
IN TWOS

WINTER BUD

ROOT

ROOT

(a)

(b)

FINELY DIVIDED
LEAVES

FORKED LEAVES

WINTER BUD

ROOT

NO ROOTS

(c)

(d)

Fig. 3. (a) *Canadian pondweed;* (b) *water starwort;*
(c) *water milfoil;* (d) *hornwort*

16

Hornwort. If you glance quickly at hornwort, you may at first think that it is milfoil, but, if you look carefully at it, you will see that the bristle-like leaves grow out of the stem in whorls of six. Each leaf is divided three or four times into forks. This plant has no roots (Fig. 3*d*). During the summer you may see very tiny flowers in the AXILS of the stem (i.e. where the leaf joins the stem). In the autumn, winter buds are formed. You may anchor one end of a piece of this weed under a stone in your aquarium.

Water starwort. The leaves of starwort grow in pairs and are submerged. The top few leaves grow very close together, and form a green star on the surface of the water, hence its name starwort. Tiny green flowers may grow above the water in the summer. Hair-like adventitious roots grow out of the stem, and may enter the mud in a pond. If you float pieces of starwort in your aquarium, you will soon see roots growing out of the stem (Fig. 3*b*).

Partially submerged plants

These plants are rooted fast in the mud. The flowers are usually conspicuous and grow above the water.

Water crowfoot. The stem of water crowfoot either floats in the water or runs through the mud, sending out roots from the NODES, i.e. from where the leaves grow out of the stem. It has two kinds of leaves. The leaves which are submerged are very finely divided, and, like milfoil, look very attractive in an aquarium. This plant is often found in running water, and the water runs easily between these finely divided leaves without tearing them. Some leaves lie on the surface of the water and have three broad lobes (Fig. 1). The flowers look like buttercups with white petals, and they can be seen

17

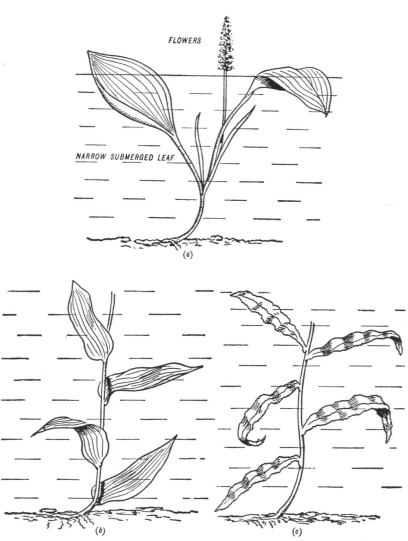

FLOWERS

NARROW SUBMERGED LEAF

(a)

(b)   (c)

Fig. 4. (a) *Broad pondweed partially submerged;* (b) *long pondweed submerged;*
(c) *shining pondweed submerged*

18

throughout the summer on the surface of the water. Place a piece of this plant in an aquarium with one end under a stone, and roots will soon grow.

Pondweed. A number of plants which have the name pond-weed are very commonly found in fresh water. They all have roots which grow for many years in the mud, sending up branches into the water. The flowers grow close together on a stalk which grows out of the water. In the broad pondweed (Fig. 4*a*), most of the leaves float on top of the water. These leaves are wide and have long stalks. The submerged leaves are cylindrical and stalked, or they may be reduced to a mere leaf stalk. The shining pondweed (Fig. 4*c*) and the long pondweed (Fig 4*b*) are totally submerged and have leaves which are long and wide but very thin, which grow, without stalks, alternately down the stem. The leaves of the shining pondweed are longer and thinner than those of the long pondweed and have wavy edges.

Water plantain. This plant has two different types of leaves. The leaves which grow above the water are rather similar to the floating leaves of the broad pondweed, while those which are submerged are long and narrow. The small pink flowers grow on a stem which may grow two or three feet high above the water (Fig. 1). The root of this plant must be set in shingle or sand in your aquarium.

There are many plants which are very common in fresh water which have not been mentioned in this book. Some of these plants, such as bur-weed, arrowhead, water parsnip and brooklime, will all grow in an aquarium.

19

CHAPTER 2

SMALL ANIMALS FOUND IN FRESH WATER

Your aquaria are now ready to house any animals that you may find in fresh water. If you have a pond or a canal near to your home or to your school, spend some of your leisure time fishing there. You will find many interesting creatures living in a pond. Put some water and water weed into your fishing can before you begin to fish, and do not put too many animals in one can. In this chapter you will learn that some animals eat other animals. Try to recognize these fierce creatures, and put them into separate jars, or else they will eat many of your specimens before you arrive at school.

The amoeba

The only way to find an amoeba is to bring a little mud back with you from a pond, and then examine it under a microscope using a well-slide so that it will not be smashed. If you are lucky you may find an amoeba. It is only one-fiftieth of an

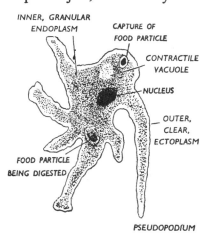

Fig. 5. *Amoeba*

20

inch big. It is made up of a soft, colourless jelly, called
PROTOPLASM, which is always changing its shape. The protoplasm is clear on the outside and forms the ECTOPLASM. The
inner protoplasm is granular and is called ENDOPLASM. You
will see a small spot which is different from the rest of the
protoplasm. This spot is called the NUCLEUS.

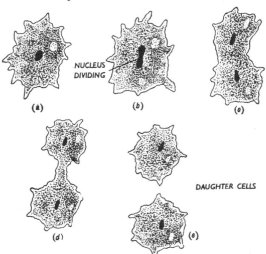

Fig. 6. *Amoeba dividing into two daughter amoebae*

The amoeba has no limbs with which to move, the protoplasm simply streams out forming a false foot or PSEUDO-
PODIUM and the remainder of the protoplasm flows after it.
It you look at an amoeba under a microscope you will see
the protoplasm moving.

The amoeba eats plants or animals that are smaller than
itself. When the food is almost touching the amoeba the
protoplasm of the amoeba flows all round it. So the food
becomes surrounded by a drop of water, within the animal.
This is called a FOOD VACUOLE (Fig. 5). A liquid from the

protoplasm containing substances called ENZYMES passes into this drop of water and digests the plant, that is, the plant is changed into a substance which will dissolve in water. The digested food passes into the protoplasm, whilst the food which has not been digested is left behind as the animal flows away from it. Surplus water does not pass out of the animal with the waste food. It collects in one spot called a CON-TRACTILE VACUOLE. This vacuole contracts at regular intervals to get rid of the water.

As the amoeba is such a tiny animal, the oxygen, dissolved in the water, passes into its body, and carbon dioxide passes out again, over the whole surface of the animal.

When an amoeba becomes too large, it divides into two (Fig. 6). First the nucleus divides into two, and then the whole animal divides. Each daughter amoeba has half of the nucleus. No parent is left to grow old and die. If the pond dries up, the amoeba becomes round and forms a thick wall or CYST round itself to prevent evaporation of water. When there is water in the pond, the cyst bursts and many small amoebae, which have formed inside it, escape.

The hydra

Hydra are only about one-quarter of an inch in length, so it is impossible for you to see them in a pond. Place some pond weed in water in a glass jar and leave it for a short time, and then you may see hydra clinging to the weeds (Fig. 7). They are either brown or green in colour.

The body is similar to a milk bottle in shape, and, like the bottle, has an opening at one end only. This opening, the MOUTH, is slightly raised and is surrounded by six to ten fine hollow threads, the TENTACLES. The mouth leads into a hollow sac called the GUT CAVITY. The amoeba consists of one cell only, but the body of a hydra consists of many cells which are arranged in two layers to form the BODY WALL. The outer layer or ECTODERM is separated from the inner layer or ENDODERM by a jelly-like substance, the MESOGLOEA. The body wall encloses the gut cavity. The end of the animal opposite to the mouth is called the FOOT. The foot gives out a sticky substance which fastens the animal to the weed.

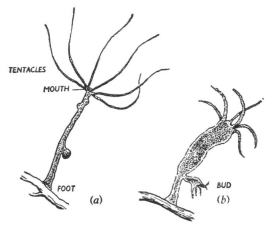

Fig. 7. *Hydra.* (a) *Fully extended;* (b) *partially contracted*

If the hydra is disturbed in any way, it immediately withdraws its tentacles, and the body becomes shortened (Fig. 7). Later the body lengthens again and the tentacles spread out. This can be seen with a hand lens, or under a microscope if the hydra is placed in water in a watch glass. The hydra is able to move from place to place by gliding slowly along on its foot, or by walking with its tentacles and its foot (Fig. 9).

Hydra eat water fleas (Fig. 16) and other small animals. So if you wish to keep hydra alive, you must pour pond water which contains these animals into your aquarium. The hydra

23

catches its food with its tentacles. There are special cells on the tentacles, called THREAD CELLS. Each cell encloses a coiled thread. A small 'trigger' projects from each cell (Fig. 8 b). When a tiny animal touches the 'trigger', out shoots the thread, which is sticky and poisonous. Some threads are not poisonous but just wrap round the animal, and others are adhesive. The tentacles close over the prey, which is drawn

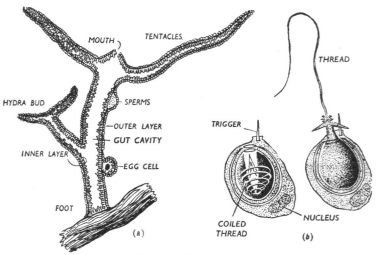

Fig. 8. (a) *Section of hydra;* (b) *thread cells*

into the mouth, and then passed to the gut cavity where it is digested. Some cells in the endoderm secrete digestive juices which begin to digest the prey. Other cells give out pseudopodia which enclose and later digest the food as the amoeba does. Digested food is passed from cell to cell throughout the animal. Some endoderm cells have hair-like projections which wave about in the gut cavity and keep the liquid moving. Undigested food has to be passed out through

24

the mouth again. Thread cells die when they have been used, and are replaced by new ones which are formed from small cells in the ectoderm.

The animal takes in oxygen, which is dissolved in the water, and gives out carbon dioxide over the whole surface of its body.

If you watch hydra closely in your aquarium, you will find that they often have lumps on their bodies. The larger lumps, which usually grow during the summer, are BUDS (Fig. 7b).

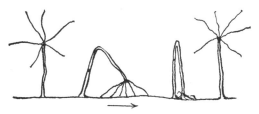

Fig. 9. *Hydra walking*

At the end of each bud a mouth and tentacles are formed. A wall then grows across the base, cutting the bud off from the parent. The young hydra moves away and starts life by itself.

The hydra can reproduce in another way. Sometimes small lumps grow on the sides of the hydra, which do not form buds. In those nearer the foot, one large cell grows. This is called the EGG. In the lumps nearer to the mouth many very tiny cells are formed. Each tiny cell has a 'head', which contains a nucleus, and a long 'tail'. These cells are called SPERMS, and when the lump in which they are formed bursts, they swim about in the water by means of their tails. The lump containing the ripe egg also bursts, exposing the egg. If a sperm touches an egg it joins with it, and then the

25

egg can grow into a new hydra. We say that the sperms FERTILIZE the eggs. Only one sperm can fertilize one egg. The fertilized egg begins to grow and is surrounded by a horny envelope or CYST. It then drops from the parent. The egg remains in the pond throughout the winter, it emerges from the cyst and grows into a new hydra the following spring.

If a hydra is cut into two, each part can grow again into a complete animal.

The horse leech

Leeches are worm-like animals which live in fresh water. You will often find them in the mud in or around a pond. If you find a leech, remember to put a cover over your jar or aquarium, as leeches often climb out of the water.

A full-grown horse leech may be four inches long. It has a flattened, dark reddish brown body which is made up of a number of rings or SEGMENTS. At each end of the body there is a SUCKER, with which it clings to any object. The MOUTH is in the middle of the front sucker, which is much smaller than the hinder sucker (Fig. 10). Leeches have very small EYES, which are so small that you may not be able to see them.

Horse leeches can swim through the water by throwing their bodies about in a wave-like manner. They can also move by means of their two suckers, grasping some plant or stone, first with one sucker and then with the other sucker.

Leeches suck the blood of other animals, after first piercing the animal's skin with their sharp jaws. After they have sucked as much blood as they want, they swim away from their prey and may not require any more food for several days. You can feed the leech in your aquarium by dropping a very large earthworm into the water. The leech may or may not kill the earthworm. Remove the earthworm from your aquarium as soon as the leech has left it.

The oxygen which is dissolved in the water passes through the skin of a leech into its blood, and carbon dioxide passes from the blood out into the water.

Fig. 10. *Horse leech*

All leeches can lay eggs. Several eggs are laid in a transparent case or COCOON, which is usually left in the damp earth round a pond. The eggs hatch into small animals like their parents.

The pond snail

When you are fishing you will find several kinds of snails. Fig. 11 shows one of the very common snails, which is called a pond snail. Nearly all snails have a spiral shell, but trumpet snails have flat shells.

The SHELL of the great pond snail is very pointed and has six or seven turns to its spiral when it is fully grown. The oldest part of the shell is at the top of the spiral. As the snail grows, more layers are added to the edge of the shell. Look at a snail shell and you will see these growth lines. The shell protects the rest of the snail's body, which is very soft.

Watch a great pond snail in your aquarium. You will see it slowly gliding along on its FOOT. From the front end of the foot, near to the mouth, a slimy substance is given out from a special gland which helps the snail to glide along easily.

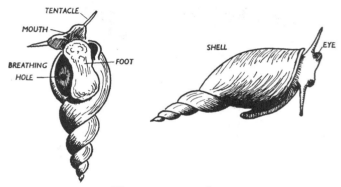

Fig. 11. *Great pond snail*

The front end of the animal is called the HEAD. Growing out of the head you will see a pair of TENTACLES or feelers, at the base of which you will see two EYES (Fig. 11). The MOUTH is situated below the feelers. Let your pond snail glide up the side of the aquarium, and watch it closely as it does so. You will see the mouth open, and an orange speck will appear. This is the TONGUE which is fastened throughout its length to the floor of the mouth and is covered with hundreds of small teeth. The tongue is like a file, and with it the snail is able to file away the leaves of the water plants, or the algae growing on the sides of the aquarium.

28

Occasionally the pond snail comes to the surface of the water to breathe. When the snail is at the surface, you will see a small hole appear between the foot and the right side of the shell. This hole is the BREATHING HOLE or SPIRACLE (Fig. 11). Oxygen from the air passes through the breathing hole into a kind of LUNG, where it enters the blood. Carbon dioxide passes out from the lung through the breathing hole. The hole is then closed and the snail can remain below the surface of the water for some time.

Fig. 12. *Eggs of pond snail in jelly on water milfoil*

All snails can lay eggs. About 20 to 30 eggs are laid in a string of jelly which is about one inch long and a quarter of an inch wide. This jelly is fastened to a weed or a stone. The young snails hatch in a few weeks, and are fully grown in about two years.

The swan mussel

Swan mussels are not always easy to find, as they usually lie embedded in the mud or stones at the bottom of any fresh water. They usually leave trails in the mud as they move along, which may help you to find them. Like the pond snail, the swan mussel's body is protected by a shell. The pond snail has a single shell, but the shell of the swan mussel is in two parts which are hinged. The shell of a full-grown swan mussel may be five or six inches long. The oldest part of the shell is that part which is nearest to the hinge. As in the pond snail, you can see the growth lines round the shell. Near to the hinge, at each end of the shell, there is a very strong muscle which is attached to each half of the shell.

These muscles can open or close the shell. You can, of course, only see these muscles if you open the shell.

Swan mussels are not very active and often remain in one place for a long time. They can move if they wish to do so. Sometimes, from the front end of the shell, you will see a cream, fleshy lump pushed out (Fig. 13). This is the FOOT, which is very muscular. The foot stretches out and then contracts, dragging the shell after it.

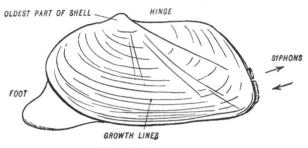

Fig. 13. *Swan mussel*

Watch a swan mussel without disturbing it, and you will see that the shell is slightly open all the way round. (If you disturb the mussel, it will close its shell.) At the hind end of the shell you will see two holes or SIPHONS, a smaller one near to the hinge, and a larger one next to it, which is fringed with hair-like projections. Water, containing oxygen and particles of food, enters through the larger siphon, and water containing carbon dioxide and waste matter comes out through the smaller siphon. If you keep swan mussels in your aquarium, you must use pond water, as it contains the microscopic animals and plants which the mussels eat. The MOUTH of the mussel is inside the shell, just above the front end of the foot.

Open the shell of a swan mussel and you will see, immediately inside each half of the shell, two reddish fleshy

30

flaps which lie on top of one another. These are the GILLS. Water which enters through the larger siphon passes over the gills. This water contains dissolved oxygen which passes into the blood in the gills. Carbon dioxide passes out of the blood into the water.

In hydra, horse leech and pond snails both male and female organs are present in the same animal. These animals are said to be HERMAPHRODITE. The swan mussel, however,

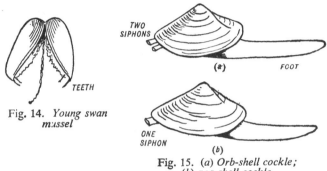

Fig. 14. *Young swan mussel*

TWO SIPHONS

(*a*)

FOOT

ONE SIPHON

(*b*)

Fig. 15. (*a*) *Orb-shell cockle;* (*b*) *pea-shell cockle*

is UNISEXUAL as the male and female organs are in different animals. The mother swan mussel has a fatter shell than the father. If you can find a female swan mussel during the autumn or winter, open the shell. You will see that the two outer gills are very fat and are filled with a brownish soft mass. Put a little of this under a microscope and you will see that it is made up of thousands of small animals with hinged shells (Fig. 14). These are small swan mussels which have hatched from the eggs which the swan mussel laid during the summer. When the eggs are laid, they pass into the outer gills which are hollow, where they hatch and remain until the following spring. In the spring the young swan mussels escape, and fasten themselves to the skin of a fish with their teeth. A hard case forms round it to protect it which looks like a white lump on the fish. Here it stays for three months, feeding on the fish. After three months the tiny mussel drops off the fish, having grown its proper shell. You can easily follow this life history if you obtain female mussels in the autumn, and keep them in a tank with sticklebacks during the following spring and summer.

The orb-shell cockle and the pea-shell cockle are two very small animals which, like the swan mussel, have double shells (Fig. 15). These animals are commonly found in fresh water, and are a quarter to half an inch long.

Water fleas

Water fleas are very important little animals to have in your aquaria, as so many other animals eat them. They are about as big as a pin's head when fully grown, and look like small white specks swimming about in the water. If you look at a water flea under the microscope, you will be able to see right through the animal, as it is transparent. See if you can find all the parts labelled in Fig. 16.

The water flea's body is flattened sideways, and all of it, except the head, is covered with a kind of SHELL which is always open down the front or VENTRAL surface. This shell cannot grow as can that of the swan mussel. As the animal grows it casts off its old shell and a new shell is secreted by its own skin. This is called MOULTING.

Look at its head and you will see a single large black EYE, and two very large FEELERS or ANTENNAE. The water flea swims or rather jumps about in the water by means of its feelers.

Inside the shell you will see five pairs of small LEGS, which are always moving. As they are inside the case they cannot be used for swimming, but have other jobs to do. Water fleas eat plants which are smaller than themselves. These plants are swept towards the MOUTH by the constant movement of the five pairs of legs.

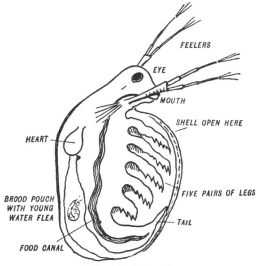

Fig. 16. *Water flea*

Water fleas breathe through their legs and through the skin on the inner side of the shell. The oxygen which is dissolved in the water passes through the legs or skin into the blood, and carbon dioxide passes out. As the legs move a fresh supply of water is brought into the shell.

Water fleas are unisexual. During the summer there are no males, the females lay unfertilized eggs into a special BROOD POUCH inside the shell (Fig. 16). Here they remain, protected by their mother's shell, until they have grown into little animals like their mother, then they escape into the water and swim away. When winter is approaching some eggs hatch into males. The female produces one or two fertilized eggs which are laid into the brood pouch which now has thick walls. When the water flea moults, the brood pouch breaks away from the animal, and is protected by the moulted shell. The eggs in the brood pouch are well protected and may remain at the bottom of the pond for months before growing into new water fleas. Water fleas usually live for several weeks or months.

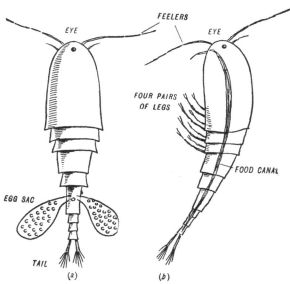

Fig. 17. *Cyclops.* (a) *Dorsal view;* (b) *side view*

Cyclops

These animals are about the same size as the water fleas, and, like them, are eaten by many other animals in the pond. They are quite easily recognized when swimming about in the water (Fig. 17).

The cyclops has a single eye and two pairs of feelers; one pair is larger than the other pair. It has four pairs of legs. Cyclops can move slowly by means of the feelers, and it can swim quickly with its four pairs of legs which are quite free.

A cyclops eats decaying matter, and is often seen feeding on the dead bodies of animals. It breathes through the surface of its body.

If you look closely at your aquarium containing the cyclops, you will see that some of them have two fairly large EGG SACS, one on each side of the body (Fig. 17a). These eggs hatch into little cyclops which are not quite like the parent. They are short and broad, and have no tail, and have not as many legs as their parents have. After growing and moulting several times they change into adult cyclops. Place a cyclops with egg sacs in a small dish containing pond water. When the egg sacs have burst, look at the water under a microscope and you will see young cyclops. Sometimes a cyclops lays larger, resting eggs which may take many weeks to hatch. These eggs are able to withstand cold and drought.

The fresh-water shrimp

Fresh-water shrimps are usually found in running water. If you wish to keep them alive in your classroom, you must keep them in a shallow dish containing weeds and decaying matter. If you keep them in deep water, it must be aerated.

The fresh-water shrimp is about half an inch long, and can be seen swimming about in a curved position on its side (Fig. 18). Its body, which is SEGMENTED, is flattened sideways. Each segment has one pair of LEGS. On its head you will see two EYES and two pairs of FEELERS or ANTENNAE.

The shrimp swims about by straightening its tail and then suddenly bending it. This action makes the shrimp move backwards rapidly. The fifth, sixth and seventh pairs of legs point backwards, and they, together with the next three pairs of forked legs, help in swimming. The last three pairs of legs are used for jumping.

On the first seven pairs of legs there are GILLS. When the shrimp is at rest you will see that the first four pairs of legs

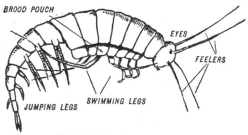

Fig. 18. *Fresh-water snrimp*

are constantly moving, bringing fresh water, containing dissolved oxygen, to the gills. The oxygen passes through the gills into the blood and carbon dioxide passes out.

Shrimps are scavengers and eat decaying matter in the pond.

The mother shrimp, like the water flea, does not lay her eggs into the water, but keeps them in a BROOD POUCH until they hatch. Plates grow out from the first seven pairs of legs, and overlap to form the brood pouch. The eggs remain between these plates and the body wall. While mother shrimp is carrying the eggs or young in the brood pouch, she is carried about by father shrimp who is much larger than she is.

The water slater or water louse

Water slaters are found in any still water, walking about over the stones or weeds. They are about half an inch long when fully grown. Their bodies are SEGMENTED, but they are

36

flattened from back to front and not sideways as in the shrimp.

Water slaters have two pairs of FEELERS or ANTENNAE and a pair of EYES. Each of the first seven segments behind the head has a pair of fairly long legs with which it is able to crawl about. Water slaters do not swim. The remainder of the segments are joined together and have six pairs of legs, of which the first five pairs look like broad plates. These plates or GILLS are used for breathing. Only the sixth pair of legs can easily be seen.

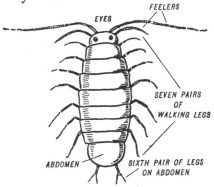

Fig. 19. *Water slater*

Water slaters eat decaying matter, and, like the fresh-water shrimps, are called scavengers.

As in the fresh-water shrimp, the mother water slater keeps her eggs in a BROOD POUCH which is made of plates which grow out of the first four pairs of legs.

Insects found in fresh water

Many insects spend part or the whole of their lives in water. We can only learn something about a few of these insects in this book, but from the few examples that we shall study, you will learn how these animals have adapted themselves to life in the water.

Carnivorous water beetle

Many beetles spend the whole of their life in water. The largest water beetle is the carnivorous water beetle, so called

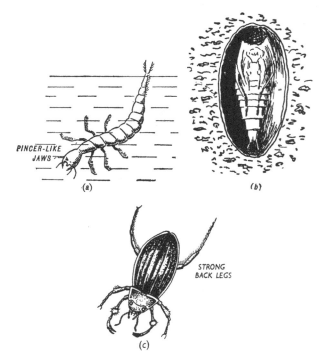

Fig. 20. *Life history of carnivorous water beetle.* (*a*) *Larva breathing;* (*b*) *pupa in soil;* (*c*) *adult swimming*

because it eats other animals that live in ponds. It is about one and a quarter inches long. These beetles are frequently found amongst the weeds in fresh water. The body of every insect is divided into three parts, the HEAD, the THORAX, which corresponds to the chest region of your body, and the ABDOMEN. On the head you will see one pair of FEELERS, and two very large COMPOUND EYES, so called because each consists of many small eyes. The thorax consists of three segments. There is a pair of legs on each segment, and a pair of wings on segments two and three. The first pair of wings is hard and horny and completely covers the top of the abdomen. The second pair of wings is hidden underneath the first pair. These wings are transparent and are used for flying.

Watch the beetle moving about in the water. You will see that it swims with its legs, particularly the last pair. These legs are long and are fringed with stiff hairs. Sometimes the beetles crawl up the water weeds, out of the water, and fly to another pond, using the second pair of wings for flying. Always cover over the top of an aquarium which contains water beetles.

The carnivorous water beetle is very fierce, eating many living, soft-bodied animals. If you keep them in an aquarium, you can feed them on tadpoles or small earthworms. Do not throw the earthworm into the water, but hold it, with forceps, in front of the water beetle's head. If the beetle is hungry, it will quickly seize the earthworm with its front legs.

Insects usually breathe through holes in the sides of their bodies, called BREATHING HOLES or SPIRACLES. Air goes through these holes into tubes called AIR TUBES or TRACHEA which branch all over the insect's body. So the oxygen passes to all parts of the body, and carbon dioxide passes from all parts of the body out through the breathing holes. In our bodies we have no air tubes. The oxygen is carried to all parts of the body in the blood. The carnivorous water beetle comes to the surface of the water to breathe. Air passes into two large breathing holes at the end of its abdomen, and more air passes into the space between the abdomen and the wings. This supply of air is used up whilst the beetle is swimming in the water..

Mother beetle has a sharp, pointed tube at the end of her body, with which she lays her eggs inside the stems of water plants. In about three weeks the eggs hatch into little animals, called LARVAE, which are nothing like their parents. Fig. 20a shows a larva in its usual attitude in the water. The larva holds on to the weeds with its six legs, or swims with its tail.

When the larva wants to breathe, it floats to the surface and pushes the tip of its abdomen out of the water. Air passes into the two breathing holes at the end of the abdomen. These holes are closed when the larva is under the water.

The larvae, like their parents, are very fierce creatures and eat any soft-bodied animals that come their way. Put some tadpoles into your aquarium containing these larvae, and watch them closely. The larva remains perfectly still, and if an unsuspecting tadpole comes near to it, it quickly grasps the tadpole with its pair of pincer-like jaws, which are hollow. Through these hollow jaws, the larva sucks the blood of its victim. The mouth has a slit-like opening which is seldom used, although some small solid particles may enter through this slit. The larva has an enormous appetite and grows very quickly. As it grows it MOULTS, because, like the water flea, its body is covered with a kind of case which cannot grow. In four or five weeks the larva is about two to three inches long. It then leaves the water and goes into the damp earth near to the pond. Here it makes a little hole, and then sheds its skin, and looks nothing like the larva or the beetle (Fig. 20b). It is now called a PUPA. A pupa does not eat, but rests whilst the larva gradually changes into a beetle inside the pupal case. When the beetle is formed, the case of the pupa splits and out comes the adult beetle, which may live for several years. The pupae which are formed at the end of the summer may not change into beetles until the following spring.

Water scorpion

Like the carnivorous water beetle, the water scorpion spends the whole of its life in the water. Water scorpions are very lazy creatures, and move very slowly by walking amongst

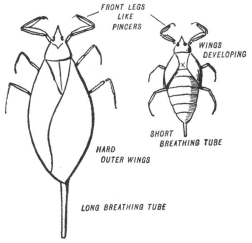

Fig. 21. *Water scorpion and nymph*

the weeds. They often remain still amongst the weeds for a long time. Like the water beetles, they have wings, but they rarely need to use them.

Fig. 21 shows you what a water scorpion looks like. It is brown in colour and about one inch long.

A water scorpion eats other small living animals that come within its reach. It seizes and holds the prey with its peculiar

41

front legs (Fig. 21). The mouth of the water scorpion is pulled out into a kind of BEAK, which is pushed into the prey. The water scorpion then sucks the blood of its victim through this beak. You can feed water scorpions on small earthworms.

The BREATHING TUBE is at the end of the abdomen. It is made up of two pieces which are usually held together to form a hollow tube. This tube is pushed out of the water, so that the air, containing oxygen, can pass into the breathing holes.

The mother scorpion lays her eggs on the water plants. The eggs hatch out into small scorpions, which are similar to their parents, but have no wings and only a short breathing tube. The wings gradually grow and the breathing tube gets longer at each moult. The young scorpion is called a NYMPH.

The water boatman or Corixa

There are two animals which you may find in a pond which are commonly called water boatmen. One swims upside-down (Notonecta) and the other one (Corixa) swims the right way up. Both animals swim by means of their long third pair of legs which are fringed with hairs.

Corixa. Look at Fig. 22*b* and you will see what a water boatman looks like. When full-grown it is about half an inch long. It has two pairs of wings. The second pair is hidden beneath the horny first pair which completely cover the abdomen. The back or DORSAL side of the body is flattened. If you keep water boatmen in an aquarium, you can often hear them hitting the sides of the glass as they swim.

Like the water scorpion, the water boatman has a BEAK through which it sucks up its food. This water boatman eats decaying matter.

The water boatman has to swim to the surface of the water to breathe. It pushes the tip of its abdomen out of the water, and air passes into a space between the wings and the abdomen. From this space, oxygen can pass into the BREATHING HOLES in its abdomen, even when the animal is under the water.

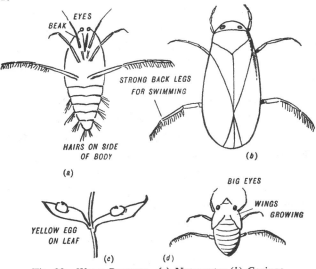

Fig. 22. *Water Boatmen.* (*a*) *Notonecta;* (*b*) *Corixa;* (*c*) *Corixa eggs;* (*d*) *Corixa nymph*

The water boatman's eggs are pale yellow, and about as big as a pin's head. They are laid singly on water plants during the spring or summer. They hatch into small water boatmen which look like their parents, but they have not any wings. The wings gradually develop as the animal grows and moults. Small water boatmen eat the same kind of food as their parents. They breathe through their skins.

Notonecta. Notonecta is larger than Corixa, and always swims upside-down (Fig. 22*a*). The dorsal surface of the body is keel-shaped, and because it is always underneath it is lighter in colour than the ventral surface of its body which is always on top.

Like Corixa, it is a very strong swimmer, but as it is lighter than water it has to cling to the weeds when it is at rest, or else it would float to the surface.

It floats easily to the surface of the water to renew the air which is carried amongst the hairs at the side of its abdomen.

The life history is similar to that of Corixa. The eggs, however, are partly hidden inside the water plant, and do not rest on the top of the leaf.

Notonecta sucks the blood of any other animal in the pond, and even attack animals much larger than itself.

The water beetle, water scorpion, Corixa and Notonecta all spend the whole of their lives in the water, although, when fully grown, they have wings with which they could fly from one pond to another. The following insects spend the early part of their lives in water, but when fully grown fly about in the air. If you keep these animals in school, you must have plants in your aquarium which will grow above the water, so that they can climb out of the water.

The caddis-fly

When you are fishing, push your net amongst the weeds, and you will probably find small cases which are made of leaves, twigs, shells or grit. If you look inside one of these cases, you will see an animal moving about. This animal is a young caddis-fly. Put one of these animals into your aquarium and you will see that it pushes its head, thorax and six legs out of the case. All these parts of the body are protected by a hard covering. The soft abdomen remains inside the case.

The caddis-fly LARVA walks about over the weeds or stones by means of its six legs, and drags its case after it. If you carefully remove the larva from its case, you will see that there is a pair of hooks at the end of the abdomen, which grasp the case when the larva is walking along. There are

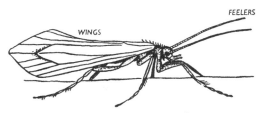

Fig. 23. *Adult caddis-fly*

also three fleshy lumps on the first segment of the abdomen which push against the case when the animal's head is out of the case (Fig. 24 *b*).

Caddis-fly larvae eat water plants.

Look at a larva, without its case, and compare it with Fig. 24 *b*. The abdomen is made up of nine segments. From the sides of each segment, except the first and the last, you will see hair-like outgrowths. These are the GILLS through which the animal breathes. The larva wriggles about inside the case, so causing a stream of water to pass through the case. The oxygen from the water passes into the AIR TUBES in the gills and carbon dioxide passes out.

We have already learned that there are many different kinds of caddis-fly larval cases. See how many you can find. When making its case, the larva cements the pieces together with silk which can be spun from a tube just below its mouth. More pieces are added as the animal gets larger.

When the larva is fully grown it closes the front of its case

45

with a silken mesh, and remains amongst the stones or weeds. It then changes into a PUPA (Fig. 24*d*). The pupa continues to wriggle its body inside the case, and to breathe through gills, but it does not eat. In two or three weeks it is ready to change into a caddis-fly. The pupa has strong jaws and is able to bite its way out of the case. It then swims to the surface of the water and climbs up a water plant out of the water. The skin of the pupa splits and the caddis-fly emerges.

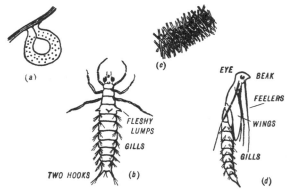

Fig. 24. *Life history of caddis-fly. (a) Eggs in jelly attached to weed; (b) larva; (c) larval case; (d) pupa*

The caddis-fly (Fig. 23) is often seen flying near to a pond. It looks like a brownish moth. When at rest, its wings slope down from the back to the sides rather like the roof of a house.

The dragon-fly

During the summer and autumn you may see large and small dragon-flies flying about near to a pool. They are very beautiful insects, with four large transparent wings and two very large compound eyes. Try to catch a dragon-fly, it will not sting you, and compare it with Fig. 25.

The dragon-fly has six legs with which it clings to any object. Dragon-flies do not walk, they only fly from place to place, catching their food as they do so. They eat flies, gnats and mosquitoes.

Fig. 25. *Dragon-fly*

We have already learned that insects which live in the air breathe through BREATHING HOLES which are along the sides of the body.

Long-bodied dragon-flies lay their eggs inside the stems of water plants, whilst short-bodied dragon-flies scatter their eggs on the water. The eggs hatch in about a month, or they may not hatch until the following spring. Young dragon-flies live entirely in the water for one to three years and are called NYMPHS. Their wings gradually appear.

Nymphs eat living animals that are smaller than themselves. They usually lie in wait for their prey, which they catch in a peculiar manner. From underneath the head, an arm-like thing projects which can be folded under the head, when not

47

in use. This arm is called the MASK. At the end of the mask there are two curved claws. When an animal comes near to the nymph, the mask is shot out very quickly, and the prey is held by the curved claws. The arm is then pulled back rapidly, so that the prey is held against the mouth.

The nymphs breathe in a strange way. At the end of the abdomen there is an opening called the ANUS which is the end of the food canal. If we could look at that part of the food canal which is just inside the anus, we should see small

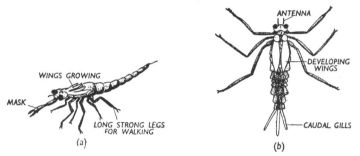

Fig. 26. *Dragon-fly nymphs.* (a) *Large kind;* (b) *small kind*

lumps projecting from the walls. These are GILLS. The nymph is able to pump water in and out of the anus, so that the gills are constantly washed with fresh water. Oxygen passes into the gills, and carbon dioxide passes out. The nymphs of some of the smaller dragon-flies have three tails at the end of the abdomen (Fig. 26b) which contain air tubes and act as gills.

When the nymphs are full grown, they climb out of the water. Their skin splits along the back of the thorax, and the dragon-fly emerges. It remains on the weed whilst its wings expand and dry. Then the dragon-fly flies away.

The life history of a dragon-fly can be watched in the

classroom. Feed very small nymphs on water fleas and larger nymphs on any other small animals, some will eat small tadpoles. You must have weeds in the tank which will grow above the water, so that the nymph can climb out of the water.

The gnat

We have all seen gnats and midges flying about in swarms on a warm day. Catch a gnat, and look at it, and compare it with Fig. 27. You will notice that it has only two WINGS. This is a characteristic of all flies. When at rest, a gnat usually stands on its front four legs, whilst the back two legs are held up above the abdomen. Look closely at the feelers of several gnats, and you will see that some of them have BUSHY FEELERS. These gnats are father gnats and only suck the nectar out of flowers for their food. Mother gnats do not have bushy feelers, and are very harmful because they suck the blood out of other animals. When feeding, mother gnat first pricks the skin and then puts a liquid into the wound, which prevents the blood from clotting. She then sucks up the blood. It is the liquid which the gnat puts into your blood which makes your gnat bites irritate.

Gnats live in water until they are full grown. Mother gnat lays about two or three hundred eggs in any stagnant water. The eggs are stuck together to form a kind of raft. After two or three days the end of the egg, which is in the water, splits and the young gnat swims away. The life history of a gnat is similar to that of the carnivorous water beetle. The EGGS hatch into LARVAE which are not at all like their parents. These larvae change into PUPAE which swim and breathe but do not eat. Inside the pupal case the larva changes into a gnat.

If you look at stagnant water, you may see larvae and pupae resting on the top of the water. They will dart away very quickly if you disturb the water.

The larva is about a quarter of an inch long (Fig. 27). It eats tiny plants and animals, which it collects on a pair of mouth brushes which are constantly moving about.

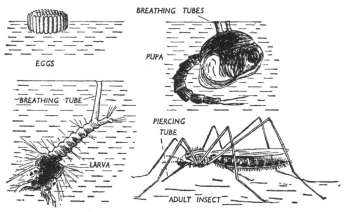

Fig. 27. *Life history of a gnat*

Look at a larva and compare it with Fig. 27. You will see a BREATHING TUBE projecting from the eighth segment of the abdomen. The end of the breathing tube is closed by five tiny flaps when the larva is under the water. When the larva comes to the surface of the water to breathe, the valve opens and oxygen passes down the tube into the insect's body, and carbon dioxide passes out of the animal.

The pupa can easily be recognized. It looks top-heavy, as the head and thorax are very large compared with the size of the abdomen (Fig. 27). The pupa swims about by wriggling its body and its two tail flaps. It does not eat, but it has to breathe. Look at a pupa and you will see two breathing tubes growing from the head end of the animal. When the pupa wishes to breathe, it swims to the surface, and pushes the two breathing tubes out of the water. When the pupa is swimming in the water, hairs prevent the water from entering the breathing tubes.

When the larva has changed into a gnat, inside the pupal case, the pupa swims to the surface of the water. Its skin splits along the back of the thorax and the gnat emerges.

CHAPTER 3

LARGER ANIMALS FOUND IN
FRESH WATER

All the animals that we studied in Chapter 2 belong to one group of animals which do not have any bones. In this chapter we shall learn something about two groups of animals that do have bones.

Fish

It is very easy to keep fish alive in an aquarium if you look after them properly. If the fish were found in running water, you must aerate your tank, or else you must keep the fish in shallow water with plenty of water weed. Remember that large fish need a big aquarium.

The stickleback

The stickleback (Fig. 28) lives quite happily in an aquarium. This fish can be recognized by the three spines on its back. A large stickleback may be two and a half inches long. You must feed them two or three times a week on small pond animals. Large sticklebacks will eat tadpoles or small earthworms, small ones will eat water fleas.

If you watch a stickleback swimming about in the water, you will notice that it is adapted for life in water. A fish's body is STREAMLINED, or tapering at each end, so that it can move easily through the water. The body is covered with

SCALES. Look closely at the scales and you will see that each scale overlaps the one behind it. The scales on the upper surface are darker than those underneath, to blend with the mud when seen from above. The lighter scales under the body blend with the light when seen from below.

Compare the fish's EYES with your own eyes. The fish has no eyelids, and so it is not able to blink or to close its eyes. Our eyes must be protected from bright light, and from dust. In a pond, the light is not very bright, and there is not any dust, and so the fish does not need eyelids. You will now understand why you must not put an aquarium, which contains fish, close to a window.

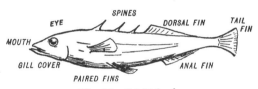

Fig. 28. *Stickleback*

Watch a fish moving about in the water, and try to find out how it swims. You will notice that the fish swims about by means of its TAIL and its FINS. There are two pairs of fins which correspond to our arms and legs, a dorsal fin on its back, an anal fin on the ventral side of its body, and a tail fin. The fish is pushed forwards through the water by the movement of its tail. The fins keep the fish in an upright position and guide the fish through the water. Inside the animal there is an AIR BLADDER which opens into the gut. The amount of air in the bladder can be varied to enable the fish to swim easily at different depths in the water.

You will notice that the fish is constantly opening and shutting its mouth, and at the same time a flap behind each eye is lifted up and lowered again. The fish is BREATHING. Beneath each flap, or GILL COVER, there are four GILLS. Each gill looks something like a curved comb with teeth on both sides. The large 'teeth' are the gills, the smaller ones to the inside are gill rakes which prevent solid matter from clogging the gills. The gills of a stickleback are too small to look at, so look at the gills of a dead herring. The gills contain blood. Oxygen passes from the water through the gill into the blood, and carbon dioxide passes out from the blood into the water. When the fish opens its mouth, water goes in. The fish then shuts its mouth and the water is forced over the gills and out as the gill cover is raised. The water does not go into the fish's stomach, as two flaps of skin, one from the roof and one from the floor of the throat, block the way.

Many fish, like the stickleback, do not have teeth, and so they swallow their prey whole. Other fish, like the very large conger eel which is found in the sea, have teeth and are able to bite.

Fish are not, as a rule, good parents. They usually lay hundreds, thousands, or, as in the herring, millions of eggs which they do not look after. Many of the young ones are eaten by other animals, but as there are so many of them, there is a good chance for some of them to survive. There are some good parents amongst the fish, that lay only a few eggs, which are taken great care of. The stickleback is a good parent. If you catch a father stickleback (or, as he is often called, a red butcher), which has a red chest and beautiful blue eyes, and one or two large, fat, dull-looking mother sticklebacks, and put them into a large aquarium which has some sand as well as stones on the bottom, you may get some young sticklebacks. The father stickleback either makes a nest of weed, or makes a hole in the sand. One or more mother

54

sticklebacks will lay their eggs in the nest. The father then sheds sperms on the eggs to fertilize them, and if the nest is in the sand, he covers the eggs with weed. He then remains on guard until the eggs hatch, chasing away any animal that comes near. He will even drive away the mothers who laid the eggs. Father stickleback guards the baby sticklebacks for some time.

The eel

The eel and the salmon are both interesting fish because they live partly in fresh and partly in salt water.

The eels, which live in all the streams and rivers of Europe, travel down to the sea when they are about six or seven years old. They travel hundreds of miles to the Sargasso Sea, which is near to the West Indies. This sea is a floating mass of sea-weed. Here the eels lay their eggs, or SPAWN, but they never return. The little eels find their way back to the European rivers taking two or three years to do so, and travelling near the surface of the sea.

Eels are easily kept in a large aquarium. Place large stones in the aquarium, under which the eels may hide. They will eat earthworms. Do not forget that when the eel is old enough, it will feel the urge to go to the sea, and consequently will jump out of the aquarium.

The salmon

Large salmon are four to five feet long. They are found living in the sea, on both sides of the North Atlantic and North Pacific Oceans. When the salmon is ready to lay its eggs, it travels up the rivers, often jumping waterfalls on its journey. Eggs are laid from September to January. The young salmon remain in the rivers for about two years, and then go back to the sea where they feed and grow rapidly. After about two years in the sea, they return to the rivers to spawn.

Other fish

Unfortunately, we cannot mention in this book all the fish that are found in fresh water. Bullheads, like sticklebacks, are easily kept in an aquarium, with very little trouble. They

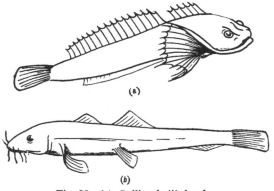

Fig. 29. (*a*) *Bullhead*; (*b*) *loach*

like shallow water, and stones under which they can hide. They will eat small earthworms. Stone loach are also commonly found under stones in streams. Minnows, which are something like sticklebacks without spines, are also frequently found. Stone loach and minnows will not live easily in captivity unless they are kept in a well-aerated aquarium.

Amphibians

If we are fishing during the spring, we always catch many tadpoles, which, as you probably know, are young frogs or toads. Frogs, toads and newts belong to another group of animals which have bones. These animals are called AMPHIBIANS. This name comes from a Greek word 'amphibios' which means 'leading a double life'. All these animals live like fish for the first part of their life, and live on land when they are full-grown.

Frogs

Put some frog's eggs, or frog SPAWN as it is called, into an aquarium containing water plants, and watch it gradually changing into frogs (Fig. 30). Each black egg is surrounded by jelly which protects it. The egg contains yolk which will feed the young tadpole until it is able to obtain its own food. The white side of the egg contains more yolk than the black side. After a few days, the egg begins to change shape (Fig. 30c, d). The head end can be distinguished from the tail end. At this stage, which is about two weeks after the spawn has been laid, the baby frogs or TADPOLES come out of the jelly. Look at one of these tadpoles with a lens and you will see a SUCKER on the underside of the head, with which it clings to the jelly (Fig. 30f). You will see three EXTERNAL GILLS which have grown on either side of the head. Until now, the tiny tadpole has taken in oxygen through its skin. Tiny blood vessels pass into the external gills. If you place one of these tadpoles in a spot of water in a watch glass and look at it under a microscope you will be able to see the blood moving in the gills. Oxygen passes out of the water into the blood, and carbon dioxide passes from the blood into the water. The tails of the tiny tadpoles grow and they wriggle to the water plants, clinging to them with their suckers (Fig. 30g). The tadpole's mouth now opens. It is surrounded with HORNY JAWS which enable it to rasp away leaves. The sucker gradually disappears and the EYES open (Fig. 30h). Four slits appear on each side of the head and internal gills grow. The external gills get smaller and the tadpole breathes through its internal gills. Folds of skin

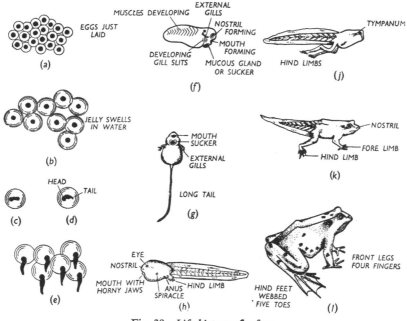

Fig. 30. *Life history of a frog*

grow over the slits on each side of the head, the one on the right side completely covers the slits. A spout-like opening, the SPIRACLE, remains on the left side. The tadpole breathes as the fish does, but the water passes out through the one spiracle after passing over the gills (Fig. 30h), and not through the two openings as in the fish. The tadpoles now begin to eat meat. They will eat animals, dead or alive, and will even eat each other. The body and tail grow much bigger. LUNGS develop inside the thorax and NOSTRILS appear on the head. The tadpole ceases to breathe through its internal gills, which gradually disappear, and breathes through its nostrils and lungs. It now has to come to the surface of the water to breathe. This it does when it is about three months old. Meanwhile the limbs grow. The two HIND LEGS appear first (Fig. 30j), and later the two FORE LEGS. The left fore leg grows through the spiracle, and the right one through the skin covering the gill slits (Fig. 30k). At this stage the tadpoles should be taken out of deep water and put into a vivarium containing a small pool (Fig. 33).

The tadpole now ceases to feed. It casts off its larval skin and its horny jaws. The tail shortens and the legs lengthen. The back legs of a frog are very long and strong for jumping, and have five long toes which are webbed for swimming. The front legs are shorter and have four fingers which are not webbed.

A frog breathes partly through its skin, which is always moist and is well supplied with blood, and partly through its nostrils. If you watch a frog you will see its throat moving up and down. As the floor of the mouth goes down, air enters the mouth through the nostrils. Oxygen can pass into the blood in the mouth, as the skin lining of it is well supplied with blood vessels, or the air can pass to the lungs where oxygen can enter the blood. The frog usually breathes through its lungs when it is very active. Normally, a frog moves slowly, and often hides, waiting for its prey to come near. Its colour matches its surroundings so that it is not easily seen.

A frog catches its food in a peculiar manner. Its mouth is very wide and its tongue grows from the front of the mouth,

59

and can be stretched a long way out of its mouth. When the frog sees an insect it quickly flicks out its tongue. The insect is caught on the sticky tongue, which is quickly withdrawn into the frog's mouth. If you watch a frog eating, you will see that its eyes become flat as it swallows. The eyes bulge into the mouth and help to push down the food. Frogs eat any small animals and can be fed on earthworms.

Frogs sleep or HIBERNATE during the winter, either in a very damp place, or at the bottom of a pond.

In spring the frogs awake and go to a pond to lay their eggs. You may hear the frogs croaking to one another. They can hear one another as they have an ear drum or TYMPANUM behind each eye.

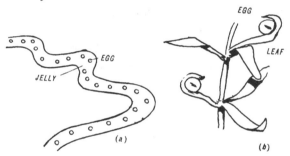

Fig. 31. (a) *Toad's eggs;* (b) *newt's eggs*

Toads

Toads are very similar to frogs, but they have a dry, warty skin, and not a smooth, wet skin like that of the frog. Toads usually walk, or hop short distances on all four legs, and so their hind legs are not as long as a frog's, but they are webbed. Toads are very much easier to keep in captivity than are frogs, and will live happily for many years if you feed them once or twice a week on earthworms. The toad pushes the large earthworms into its mouth with its front legs.

60

A toad's eggs are laid in a very long string of jelly which is wrapped round the weeds. In this string of jelly, which might be 15 feet long, the eggs look as though they are arranged in a double row, but as you pull a string of toad's eggs out of the water, the eggs then seem to be in a single row. The life history of a toad is similar to that of a frog. A toad is not full-grown until it is five years old, and it lives for several years after that. Toads hibernate in damp places, usually digging a hole for themselves in the damp earth.

Newts

Newts differ from frogs and toads because they do not lose their tails when they are full-grown. The four legs are short and are equal in size. The front legs have four fingers and the hind legs have five toes which are not webbed. The newt swims with its tail, and uses its four legs for walking on land. Newts, like frogs, have a wet skin through which they can breathe.

The crested or great water newt, which may be six inches long, and is black with bright orange marks on the under-side, will live for several years in captivity, feeding on earthworms. They often become very tame and will take an earthworm out of your hand. Although this kind of newt is called a crested newt, only father newts have a wavy crest along their backs.

Like frogs and toads, newts only live in the water when they are laying their eggs. Newts' eggs are not easy to find, as each egg is laid separately on a leaf. Mother newt curls the leaf over the egg to protect it, although the egg is partially protected by the jelly which surrounds it (Fig. 31 b). If in the spring you catch a mother and father smooth newt, which are only about three inches long, and put them into an aquarium containing Canadian pondweed, you may be

61

fortunate enough to get some eggs. The water should not be too deep, and you should have stones coming above the water, so that the newts can crawl out of the water. Cover the top of the aquarium, as newts will easily climb up the glass.

The young newt does not develop in quite the same way as the frog and the toad. The front legs appear before the back

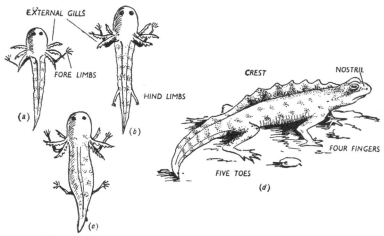

Fig. 32. *Life history of a newt. (a), (b) and (c) are stages in development; (d) father newt*

legs (Fig. 32), and the young newt retains its external gills for quite a long time. Young newts look very much like small fish as they swim about in the water, but they are recognized by their feet and external gills.

Several newts hibernate together in a damp place. They twist around one another, so that they will not dry up.

A vivarium for amphibians

If you wish to keep frogs, toads or newts in your classroom, you must make a vivarium for them. You can use an aquarium (even a cracked one) or a large box. Put soil in the bottom of the vivarium. Place a small dish in the soil, which will be a small 'pond'. Inside the dish put small stones, water weed and water, and try to camouflage the dish so that it really looks like a 'pond'. Be sure that the animals can crawl in and out of the 'pond'. Arrange some large stones inside the vivarium to look like a cave, as the animals like hiding places. Now cover the soil with small tufts of grass and moss. Cover the top of the vivarium (Fig. 33).

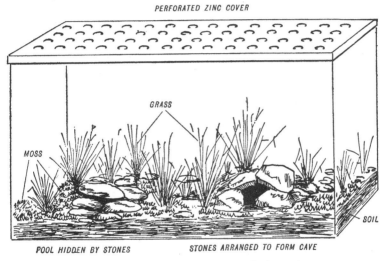

Fig. 33. *A vivarium made from a cracked aquarium*

CHAPTER 4

ANIMAL LIFE IN THE GARDEN

Many of you have a garden, or know where there is a piece of land or park where you may look for small animals. In this chapter we shall study a few animals that can be found in the garden. If you find animals that are not mentioned in this chapter, you will be able to identify them if you use the reference books that are mentioned in Appendix C.

The earthworm

If you dig in damp soil you will find earthworms. Keep your earthworms in damp soil until you are ready to look at them, because earthworms die if they become dry.

Look closely at a large earthworm and compare it with Fig. 34. It has a long cylindrical body which is pointed at both ends to enable it to tunnel through the soil. It is not easy to see which is the head end, but the earthworm usually pushes its HEAD forward when it is moving. Look at the head through a lens. You will see that the earthworm has a MOUTH but it has no nose and it has neither eyes nor ears. The back or DORSAL side is darker in colour than the lower or VENTRAL side.

The body is divided up into a number of rings or SEGMENTS. The number of segments varies in different kinds of earthworm. Rub your fingers along the earthworm's body and you will feel that it is rough on one side, the ventral side. This

is due to the presence of four pairs of little BRISTLES or
CHAETAE on each segment. Place the earthworm on a piece
of paper, and as it moves along you will hear a scraping noise
which is caused by the bristles scratching on the paper. Watch

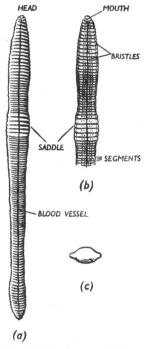

Fig. 34. *Earthworm:* (*a*) *dorsal
view;* (*b*) *ventral view showing
bristles;* (*c*) *cocoon*

Fig. 35. *Earthworm coiled up in
chamber in the soil*

the earthworm moving along and you will see that it alternately stretches and contracts its body. The bristles point
backwards and prevent the earthworm from slipping backwards as it moves along. They can be retracted as the body
moves forward.

Earthworms eat any organic matter (i.e. animal or plant matter) that they find in the soil Frequently, they eat the soil, but only the organic matter in the soil is digested. The remainder of the soil passes through the earthworm's body and is left on the surface of the ground as WORM-CASTS which look like twisted pieces of soil. Try to find some worm-casts on a lawn.

On p. 64 you read that earthworms die if they become dry. This is because they breathe through their skins. Gases can only pass through the skin if it is moist, and so if their skin dries, the earthworms cannot breathe and so they die. The skin is kept moist by a slimy substance which oozes out of the earthworm's body. Oxygen passes through the skin into the blood, which carries the oxygen to all parts of the body. Carbon dioxide is collected from all parts of the body by the blood and is carried to the skin where it is given off. The skin is well supplied with small blood vessels which you cannot see, but you will see two of the main BLOOD VESSELS, one on the upper or DORSAL side and one on the lower or VENTRAL side of the body.

Earthworms have no eyes because they spend most of their lives in the soil. They are, however, sensitive to light, and usually only come above ground when it is not too light. Earthworms can neither see nor hear, but they can feel the ground shaking if an animal is hopping or is walking close by.

All earthworms lay eggs when they are fully grown because they are hermaphrodite (see page 31). Two worms come together with their heads pointing in opposite directions. Each worm passes sperms from two holes in segment 15 into two pairs of sacs in segments nine and ten. The sperms are stored here and the worms separate. Look at your earthworm and you will see a thickened part about one-fifth of the way along the body from segments 32 to 37. This is called the SADDLE or CLITELLUM. The clitellum secretes a sticky substance which hardens to form a leathery case, which becomes loose round the earthworm. This case or COCOON is pushed forwards over the earthworm's head. As the cocoon passes over the segments behind the head, food stuff oozes from the earthworm's body into the case, eggs are laid

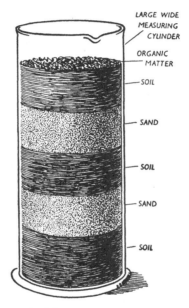

LARGE WIDE
MEASURING
CYLINDER

ORGANIC
MATTER

SOIL

SAND

SOIL

SAND

SOIL

Fig. 36. *A wormery*

into the cocoon from two holes on the fourteenth segment and sperms pass into it from segments nine and ten. When the cocoon is free the two ends seal up. The cocoons are yellowish in colour, and look something like a lemon, about one-tenth to one-quarter of an inch long (Fig. 34c). They are laid in damp places where there is plenty of decaying matter. The eggs are fertilized inside the cocoon. Only one or two worms hatch out of each cocoon. If an earthworm is accidentally cut in half, the tail end dies but the head end grows a new tail. If the weather is very cold or if it is very dry, earthworms tunnel deeply into the soil, make a little chamber for themselves, and then coil themselves up into balls (Fig. 35).

Earthworms are very useful animals in the garden as they loosen the soil, thus allowing water to drain through and air to enter the soil. Earthworms also mix up the soil, bringing the deeper subsoil to the surface. This can be seen if you set up a WORMERY (Fig. 36).

A Wormery. Put alternate layers of damp soil and damp sand into a tall, wide jar, and place some organic matter on top of the soil. Put six large worms into the jar, and cover it with brown paper, so that the earthworms will be in the dark even if they come to the edge of the jar. Leave the wormery for several weeks and you will see tunnels that the earthworms have made. You will also notice that the sand and the soil have become mixed up.

Slugs and snails

In Chapter 2 we learned a little about pond snails. There are many kinds of land snails and slugs, and they are very similar to pond snails in many ways. If you lift up plants or stones on a rockery, or look in sheltered damp corners or along hedgerows, you will find snails and slugs at any time of the year. These animals can be kept in school. Put some damp soil in the bottom of a glass jar or tank and plant grass or other plants in the soil, so that it looks like a small garden. The plants may die, but fresh plants can be planted when necessary. Keep the atmosphere in the jar damp by covering the top with glass, and leave a small space to allow air to enter (Fig. 37).

Slugs are very similar to snails, but snails have shells and slugs have not. Place a slug or a snail on a small piece of glass and look closely at it, comparing it with Fig. 38. It

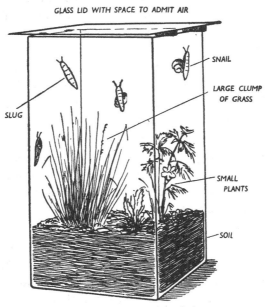

GLASS LID WITH SPACE TO ADMIT AIR

SNAIL

LARGE CLUMP OF GRASS

SLUG

SMALL PLANTS

SOIL

Fig. 37. *Vivarium for snails and slugs*

has a distinct HEAD which has two pairs of feelers or TEN-TACLES. The smaller tentacles are similar to those of the pond snail and are sensitive to touch. At the end of each of the longer tentacles there is an EYE. Do you remember where the eyes of the pond snail are? Touch the tips of the tentacles with a pencil and you will see that they are quickly withdrawn into the head.

Turn the piece of glass, on which the slug is crawling, upside-down and watch its FOOT as it glides along. You will see little 'waves' passing along the foot. These 'waves' are caused by the expansion and contraction of the muscles in the foot (Fig. 39). In both slugs and snails a slimy substance is given out from a gland at the front end of the foot, which enables the animal to glide smoothly along. This slimy substance dries after the animal has moved on, and looks like a silver trail.

Like the pond snails, slugs and snails have TONGUES similar to files, with which they can file away the leaves. They are very harmful in the garden, because they eat the young shoots of plants. You may see the tongue if the slug opens its mouth as it glides along the glass.

Watch your slug and you will see a hole appear on the right side of the body. This is the BREATHING HOLE (Fig. 39). If you hold a snail upside-down on the glass, you will see the breathing hole between the foot and the right side of the shell (Fig. 39). Air goes through the breathing hole into a bag-like lung, where oxygen enters the blood and carbon dioxide passes from the blood out through the breathing hole.

All snails and slugs lay eggs which are similar, but which vary in size from one-twelfth to one-eighth of an inch in diameter. The eggs are laid in damp soil which is seldom disturbed, such as the soil in rockeries or around the roots of plants which live for many years. The egg, which is yellowish but transparent, is surrounded with a layer of jelly (Fig. 39 b). About 20 to 30 eggs are laid at a time, in the soil. They hatch into little animals which look like their parents. Look daily at your tank containing slugs and snails. Make a note of the day when you see some eggs, and see how long they take to hatch.

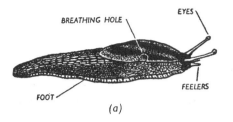

BREATHING HOLE

EYES

FEELERS

FOOT

(a)

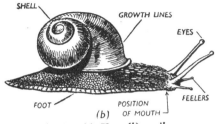

SHELL

GROWTH LINES

EYES

FEELERS

FOOT

POSITION
OF MOUTH

(b)

Fig. 38. (*a*) *Slug*; (*b*) *snail*

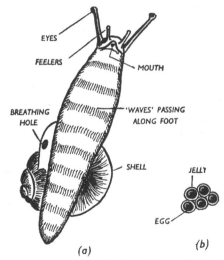

EYES

FEELERS

MOUTH

BREATHING
HOLE

'WAVES' PASSING
ALONG FOOT

SHELL

JELLY

EGG

(a)

(b)

Fig. 39. *Snail:* (*a*) *view from below;* (*b*) *eggs*

71

Many animals such as birds, toads and frogs eat slugs, which can only protect themselves by hiding amongst the plants. Snails have a SHELL into which they can withdraw. The oldest part of the shell is at the top, and as the snail grows, more layers are added to the edge of the shell. You can see these GROWTH LINES on the shell. Thrushes like to eat snails, and often have a favourite stone on to which they throw the snails in order to crack their shells.

Centipedes and millipedes

If you are digging the garden at any time of the year, you may see long, yellow, brown or black animals with many legs all along their bodies. These are centipedes and millipedes (Fig. 40). Learn to distinguish between these animals because centipedes are useful in the garden, since they eat harmful animals. Millipedes, on the other hand, are harmful and should be destroyed because they eat the roots of plants.

Both centipedes and millipedes have SEGMENTED bodies. Look closely at these animals and you will see that centipedes have one pair of LEGS to every segment, and millipedes appear to have two pairs of legs to every segment. All these animals have one pair of FEELERS, and, with the exception of some centipedes, they do not have eyes.

Centipedes and millipedes breathe in a similar manner to that in which insects breathe. They have BREATHING HOLES along each side of their bodies which lead into AIR TUBES. These air tubes carry the oxygen to all parts of the body. Carbon dioxide from the body passes along the air tubes and out through the breathing holes.

Both centipedes and millipedes lay small eggs in the ground which hatch into young animals similar to their parents. The young ones may not have the same number of segments and legs as their parents, but these develop as the young animals grow older. Their bodies are covered with a kind of shell which cannot grow, and so, as the young animal grows bigger it MOULTS.

Vivaria for soil animals. You can keep these animals, and many other kinds of soil animals, alive in a vivarium, set up as in Fig. 33. Keep a variety of small animals in the vivarium

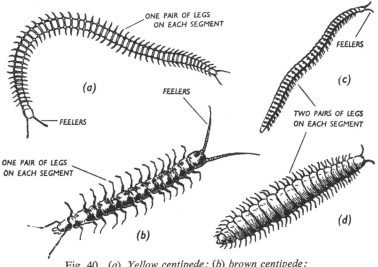

Fig. 40. *(a) Yellow centipede; (b) brown centipede;*
(c) black millipede; (d) brown millipede

for the centipedes to eat, and set many small plants that have roots, bulbs or tubers (see Chapter 7) which the millipedes will eat. In order to watch the habits of animals that live in the soil, you must set up a special vivarium as shown in Fig. 41. This vivarium must be very narrow, so that you can see the animals walking through the soil. The vivarium must be covered with dark paper as these animals do not like the light.

73

Centipedes. The most common centipede is the long yellow one (Fig. 40 *a*), which is about one to two inches long when fully grown. It is frequently called a wireworm, and so is destroyed, because wireworms are harmful. Centipedes are easily distinguished from wireworms because the former have a pair of legs to every segment, whereas the latter have only six legs (Fig. 53). Another common centipede is brown in colour, and is shorter and wider than the yellow centipede (Fig. 40 *b*).

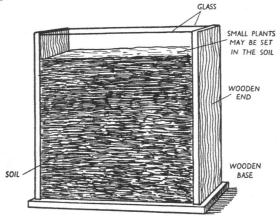

GLASS

SMALL PLANTS
MAY BE SET
IN THE SOIL

WOODEN
END

WOODEN
BASE

SOIL

Fig. 41. *Vivarium for animals that live in soil*

All centipedes are useful because they eat small animals that live in the soil. These animals may be pests. Centipedes run very quickly when looking for their prey. Their eggs are laid singly in the soil.

Millipedes are not as well known as centipedes. There are two kinds that are commonly found. One is about one and a half inches long and has a black, rounded body. Its legs are white, and are very fine and hair-like (Fig. 40 *c*). The other common millipede is light brown in colour, but is easily

74

distinguished from the brown centipede (Fig. 40 b). They lay about 60 eggs at a time in a small nest which the mother makes by gluing soil together with saliva. The eggs hatch in about 14 days. Millipedes are sluggish in habit and curl up when they are touched.

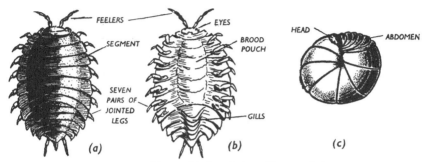

Fig. 42. *Wood-louse: (a) dorsal view; (b) ventral view;*
(c) 'pill-bug' rolled into ball

Wood-lice

Look carefully under the rocks and plants in a rockery and you will see many wood-lice, which look very similar to the water slaters that were described in Chapter 2. They are about half an inch long when fully grown (Fig. 42 a). Wood-lice eat either animal or vegetable matter. The body of the wood-louse is SEGMENTED, and it is flattened from the dorsal to the ventral side.

Look at a wood-louse and you will see that it has a pair of FEELERS or antennae. It also has one pair of EYES, but you may not be able to see them, as they are very small. Count the legs of the wood-louse. There are seven pairs of JOINTED LEGS which grow out of the first seven segments. These seven segments form the THORAX.

The ABDOMEN consists of six segments which are not joined

75

as they are in the water slater. The abdominal legs are modified for breathing. Inner plates form special GILLS which contain air tubes, whilst outer plates act as covers for these gills so that they do not dry up. If these gills become dry, the wood-louse cannot breathe, so it lives in damp places. Oxygen passes through the gills into the air tubes, and carbon dioxide passes from the air tubes out through the gills. Wood-lice will live in a vivarium (Fig. 33) if it is kept very damp.

If you look at the ventral sides of wood-lice during the summer, you will see that some of them, the mother wood-lice, have an enormous white lump between their seven pairs of legs. This is the BROOD POUCH which, as in the water slater, is formed of plates which grow out of the legs. The wood-louse carries its eggs in this pouch until they hatch. Young wood-lice look like their parents, and as they grow they MOULT. The back half of the animal moults first, and the wood-louse puts the hinder end of its body into a corner, to protect it while the new skin is growing. The wood-louse eats the moulted skin. A few days later the front end of the animal moults.

One type of wood-louse, which is commonly called a 'pill-bug' (Fig. 42c), rolls itself into a ball when it is touched.

Spiders

We are all familiar with the beautiful spiral WEBS which we see in the garden during the warmer months of the year, especially in the early morning when the dew is sparkling on them. These webs are made by the garden spider, which can be recognized by the cross on its back (Fig. 43a). Many British spiders are line weavers, and the house spider makes the cob-web, which is not sticky. Look around your garden and see how many different kinds of web you can find.

Spiders will live happily in a vivarium (Fig. 33) as long as you feed them with small flies.

Look at a garden spider. You will see that its body is divided into two parts. The HEAD and chest (or THORAX) are joined to form the first part or CEPHALOTHORAX, and the second part is the ABDOMEN (Fig. 43a). Examine the head

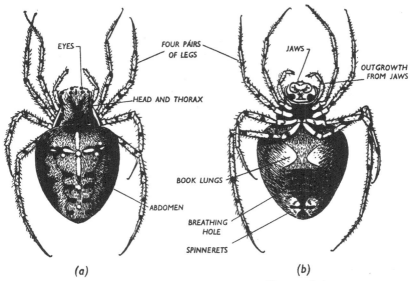

Fig. 43. *Garden spider: (a) dorsal view; (b) ventral view*

through a lens, and you will see eight small EYES. The spider has no feelers although the outgrowths from the jaw may look like feelers. You may be able to see its strong biting jaws, which have two joints (Fig. 44a). In the jaws there are poison sacs. When the spider bites its prey, the poison comes down the hollow jaws into the prey, which is then either killed or paralysed. The poison of some of the larger foreign spiders will kill a man.

77

All spiders have four pairs of jointed LEGS which grow out of the thorax. At the end of each leg there is a jointed hook, by means of which the spider can run along the finest thread (Fig. 44 b).

Hunting spiders have good eyes, because they have to catch their prey. Most English spiders, however, make webs to catch their prey.

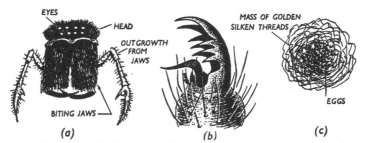

Fig. 44. (a) *Spider's head;* (b) *spider's foot;* (c) *spider's cocoon*

A spider's web

Turn your spider upside-down and look at the end of the abdomen (Fig. 43 b). You will see several small raised lumps which are the SPINNERETS. You may see only four spinnerets, but there are two more smaller ones which cannot very easily be seen. Hold the spider upside-down and place the tip of a pencil against these spinnerets, and you may be able to draw out a thread of silk. A liquid oozes out of the spinnerets which changes into a solid silken thread when it comes into contact with the air. The threads from some spinnerets are sticky, and from other spinnerets they are not sticky. The spider makes the outside threads of a web first. She presses her spinnerets against a leaf to fix the thread, and then she runs round to the next point, drawing out a thread as she goes. When she reaches the second point she pulls the thread up with her claws. When the boundary lines are completed the spider makes the spokes of the wheel. One end of the first spoke is fixed to the middle of a boundary line, and the other end is fixed to the opposite boundary line. The centre of this thread is then thickened with a mass of silk, and many more spokes are made from the boundary lines to the centre. Each thread is pulled up lightly, and any surplus silk is left in the middle. None of these threads is sticky. The spider then weaves the spiral with sticky thread. If you watch a spider making its web you will see that it places the thread with the claw on one of its back legs. When the web is finished, the spider either remains motionless in the centre of the web, or she hides in the surrounding leaves, with a thread from the web fastened to a foot. This thread is called a 'telegraph wire', because she can feel the web shaking if an insect is caught in it. The spider may bite the prey with its poisonous jaws and then suck the juices out of its body. If she is not hungry, she leaves the prey hanging by two threads, and turns it round and round, weaving a web around it so that it cannot move (Fig. 45). She then carries it away to a 'larder'. Spiders do not try to kill dangerous insects such as wasps, they either leave them alone, or set them free by biting the threads around them.

Male spiders do not make webs of their own, but steal food from the female's web. The female land spiders are often larger than the males, and sometimes eat their husbands. Male water spiders are larger than the females.

Look again at the underside of the garden spider and you may see a lighter patch on the abdomen near to the thorax, where the BOOK LUNGS are. The lungs of a spider are so called because they resemble the leaves of a book. There are 15 to 20 thin folds of skin which look like thin plates or leaves. Each leaf is hollow and contains blood which can be purified.

Just behind the book lungs there is a single BREATHING HOLE, which leads to air tubes similar to those found in the centipedes. These air tubes do not penetrate far into the abdomen.

Spiders lay their eggs in the autumn. The garden spider lays 30 to 60 tiny eggs which are held together in a mass of yellowish silken threads. These silken balls which are about three-quarters of an inch in diameter, are fastened under a ledge, window-sill, or in any protected corner (Fig. 44c). The eggs may hatch during the autumn, or they may not hatch until the following Spring. Try to find one of these COCOONS and keep it in a covered jar until the eggs hatch. Spiders MOULT as they grow bigger. You may find a whole moulted skin of a spider. Baby spiders do not eat until they have moulted the first time. Then they

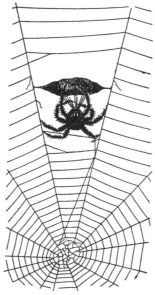

Fig. 45. *Part of a spider's web showing a spider wrapping silk round a fly*

will eat one another if they cannot find food. Usually they have dispersed before this happens. Some spiders run about dragging their firm little white cocoons after them.

Spiders may live for two to three years, and they usually HIBERNATE or sleep during the winter, because they cannot obtain food.

Spiders are useful animals in the garden, because they help to kill harmful animals. They can only protect themselves from birds, frogs, etc., by hiding.

Insects in the garden

In Chapter 2 we learned a little about some of the insects that live in ponds. There are many insects that live in the garden. You will find them in the soil, crawling about on the plants, living inside plants or flying about. Many insects are very harmful because they damage the plants, whilst other insects are useful because they eat or kill some of the harmful insects. Only a few of these insects can be described in this book. A list of books is given in Appendix C which will enable you to recognize any insects that you may find.

Butterflies and moths

Butterflies and moths are very commonly found in the garden and, at first sight, may look much alike. There are, however, two ways in which you can tell one from the other. BUTTERFLIES have a KNOB at the end of each FEELER and rest with their wings together in a vertical position. MOTHS have pointed feelers, and when a moth is at rest its wings are always folded horizontally over its body (Fig. 46). There are more varieties of moths than there are of butterflies.

See how many varieties of moths and butterflies you can find. It is interesting to make a collection of your own (see Appendix B). If you keep these insects in a large vivarium (Fig. 33) place a jar of flowers in the soil, as butterflies and moths suck the nectar out of flowers for their food. Special insect boxes (see Fig. 47) are ideal to use if you wish to study the life histories of these insects. Instructions for making home-made insect boxes are given in Appendix B.

Look clearly at a moth or a butterfly. Its body is divided into three parts, HEAD, THORAX and ABDOMEN (Fig. 46). It has two large compound EYES and one pair of FEELERS on the head, three pairs of LEGS and two pairs of WINGS grow out of the thorax. You can recognize the different kinds of butterflies and moths by their wings, which vary in colour. A few female moths, such as the Vapourer moth, do not have wings. The VEINS which you can see in the wings, strengthen the wings and contain air tubes, blood and nerves. The abdomen is SEGMENTED. Count the number of segments that you can

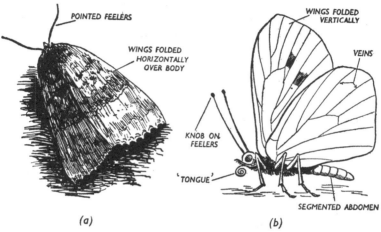

POINTED FEELERS

WINGS FOLDED HORIZONTALLY OVER BODY

WINGS FOLDED VERTICALLY

VEINS

KNOB ON FEELERS

'TONGUE'

SEGMENTED ABDOMEN

(a)

(b)

Fig. 46. (a) *Old Lady moth at rest*; (b) *Cabbage White butterfly at rest*

see. Adult butterflies and moths do not harm plants, as they only suck the nectar out of the flowers. Watch one of these insects after it has alighted on a flower, and you will see it push a very long TONGUE or PROBOSCIS into the flower. The tongue is usually coiled up under the insect's head (Fig. 46).

Butterflies and moths, like all insects, breathe throughout their lives through BREATHING HOLES or SPIRACLES which are at the sides of the body. Air is carried throughout the body in AIR TUBES or TRACHEA.

The life history of all moths and butterflies is similar. The eggs are laid in a sheltered position on, in or near to the food which the caterpillar will eat when it hatches out. Twenty to thirty eggs are laid in a cluster. The eggs hatch into tiny LARVAE which do not look like their parents. These larvae

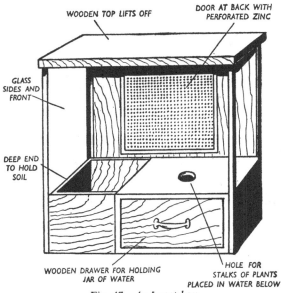

WOODEN TOP LIFTS OFF

DOOR AT BACK WITH PERFORATED ZINC

GLASS SIDES AND FRONT

DEEP END TO HOLD SOIL

WOODEN DRAWER FOR HOLDING JAR OF WATER

HOLE FOR STALKS OF PLANTS PLACED IN WATER BELOW

Fig. 47. *An insect box*

are called CATERPILLARS. Look in your garden and you will find many different kinds of caterpillar during the spring, summer and autumn. Compare them with the caterpillars that are described here (Fig. 48).

The caterpillar's body consists of a HEAD and 13 SEGMENTS. You may be able to see only 12 segments. It has six simple EYES on each side of its head and tiny feelers. Look at the head of a large caterpillar through a lens and you will see its biting JAWS (Fig). 48 *d*. It is interesting to watch a caterpillar eating a leaf because it chews the leaf, moving towards itself as it does so (Fig. 48 *c*). Most caterpillars will eat only one type of food, and will die if they are not given that food. For instance, the caterpillars of clothes moths eat wool; of the Puss moth eat willow; of the Lime Hawk moth eat lime

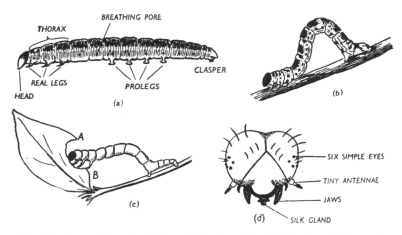

Fig. 48. *Caterpillars. (a) Cabbage White butterfly; (b) Magpie moth 'looper'; (c) caterpillar of Swallow Tail moth eating privet leaf. It always eats from A to B; (d) head of larva, front view*

leaves; of the Pea moth eat the peas in a pod; of the Peacock and Tortoiseshell butterflies eat stinging nettles; and of the Cinnabar moth eat ragwort. The Tiger moth caterpillar, on the other hand, will eat many different plants. Caterpillars are easily kept if you give them the right kind of food, and regularly clean out the jars or boxes in which they are living. If you find unknown caterpillars, give them the leaves of the plant on which they were found. Caterpillars are generally very harmful as they eat the leaves or roots of many of our garden plants and vegetables and may even burrow into a tree trunk (Goat moth).

Each caterpillar has a pair of jointed, pointed legs on each of the three thoracic segments. These are REAL LEGS which will be replaced by longer legs when it is an adult. In addition to these legs the caterpillar has either five or two pairs of FALSE LEGS or PROLEGS on its abdomen (Fig. 48). The legs are found on the sixth, seventh, eighth, ninth and last segment of the body, or on the ninth and last segments only. The pair on the last segment are called CLASPERS. Caterpillars like those of the Magpie moth, which have only two pairs of false legs, loop their backs as they walk and so are called LOOPER caterpillars (Fig. 49 b).

The BREATHING HOLES or SPIRACLES are often clearly seen along the sides of the caterpillar's body. There are nine pairs of spiracles, one pair on the first segment of the thorax, and one pair on each of the first eight segments of the abdomen.

On the underside of the head, behind the mouth, there is a small hole which leads to a SILK GLAND. A sticky fluid comes out of this hole, which becomes solid when it comes into contact with the air, and forms a SILKEN THREAD. You may see a caterpillar hanging from a tree by a silken thread.

As the caterpillar grows bigger it moults several times. When it is fully grown it finds a sheltered spot and changes into a PUPA which is called a CHRYSALIS. Before changing into a chrysalis, the caterpillar may make a protective case or COCOON around itself. The Tiger moth caterpillar makes a cocoon of silk and hairs which it pulls off its own body. If you keep them in an insect box they will climb to the top of the box to make their cocoons. Cabbage White butterfly caterpillars climb under a sheltered ledge, and fasten them-selves with one or two threads to a wall (Fig. 49). A Puss

85

moth caterpillar which lives on a willow tree makes a hard cocoon of chewed wood. Caterpillars that live in the soil (often called CUTWORMS) do not make cocoons but just pupate in the soil. Try to find as many cocoons as you can, or find the caterpillars and keep them in an insect box (Fig. 47), and watch them making their own cocoons. Caterpillars that pupate in the soil could be kept in a narrow vivarium (Fig. 41). The caterpillars that we call SILKWORMS make beautiful cocoons

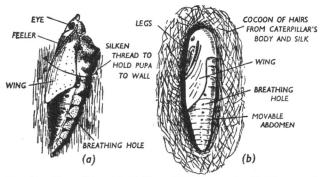

Fig. 49. *Chrysalides: (a) Cabbage White butterfly; (b) Tiger moth*

of silk which can be woven into silken material. The cater-pillars must be fed on mulberry or lettuce leaves. 'Silkworm' moth eggs may be bought from a livestock dealer.

The pupa or CHRYSALIS is the resting stage when the cater-pillar changes into a moth or a butterfly. The chrysalis can breathe through its BREATHING HOLES but it cannot eat. If you look at a chrysalis (Fig. 49) you will see the impressions of the eyes, feelers, wings and legs on the hard case. A chrysalis moves its abdomen when it is touched. The chrysalis of each butterfly and moth is easily recognized by its size, shape and colour. The chrysalis may remain asleep throughout the winter, or the adult insect may emerge the same season.

Most butterflies and moths spend the winter in the chrysalis stage, but some may spend the winter as small caterpillars which sleep in a sheltered spot (Tiger moth). The Tortoise-shell and Comma butterflies hibernate through the winter. Some moths, which are pests on fruit trees, lay eggs in the autumn which hatch the following spring.

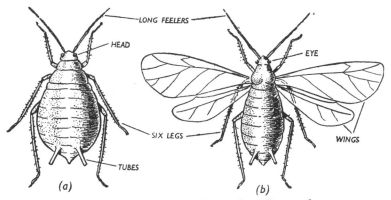

Fig. 50. *Green aphides:* (a) *wingless;* (b) *winged*

Aphides

If you look in your garden during the summer you will see masses of tiny green animals on the rose shoots. These animals are aphides but they are commonly called BLIGHT. Similar small aphides that are black are found on beans, and others, called the woolly aphides, which look like masses of cotton wool, are found on fruit trees. There are also grey aphides that live on cabbage and shrivel up the leaves, and many other kinds of aphides that live on fruit trees and even on the roots of plants. These insects will live in an insect box, if you place the shoots on which they live in a jar of water inside your insect box (Fig. 47).

87

These animals are very small so you will have to look at them through a lens. The body of an aphis is SEGMENTED and it has six LEGS (Fig. 50). It has two EYES and a pair of long FEELERS on its HEAD. Look at the underside of the head and you will see a tube-like BEAK projecting from its mouth, similar to that seen in the water scorpion and water boatman (Chapter 2). The aphis pushes this beak into the plant on which it lives, and sucks up the plant sap. This is harmful to the plant and may stunt or distort the growth of the shoot, or even kill it. If a winged aphis sucks the sap of a diseased plant, it may carry that disease to another plant.

Aphides have BREATHING HOLES along the sides of their bodies, and oxygen is carried throughout the body in air tubes. Near the end of the body there are two tubes which secrete a waxy substance. From the end of the food canal, or anus, a sweet substance called HONEYDEW is given out. Ants like to eat this honeydew and often go in search of aphides in order to get it. Ants stroke the aphides with their feelers, to hasten the secretion of honeydew.

The life-history of the aphis is very similar to that of the water flea (see page 32). During the summer there are no male aphides. The females produce unfertilized eggs which remain inside the body of the aphis until they hatch, and then they are laid. Each aphis can produce several young ones every day. Animals that are born alive like this are said to be VIVIPAROUS. When the aphides are becoming overcrowded, some of the young ones develop two pairs of WINGS (Fig. 50b), fly to another plant and start a new colony. When autumn approaches the aphides fly to different host plants which are usually weeds such as thistles and knapweeds. Here both male and female aphides are produced. The latter lay fertilized eggs which can withstand the winter and hatch

88

out the following spring. Nearly all the aphides die. In the spring these eggs hatch into little aphides just like their parents. The aphides MOULT as they grow bigger. In 8 to 18 days the young aphides are fully grown and develop into females which produce more aphides.

Aphides are very harmful in the garden and must be got rid of. They can be destroyed by spraying the plant with certain chemicals. Many insects, such as lady-birds, lace-wing flies, hover-flies and ichneumon flies, eat or kill aphides at some stage in their lives. Birds also eat aphides.

Frog-hoppers

You may not know what a frog-hopper looks like, but you must have seen 'CUCKOO SPIT' which is a frothy substance that is made by baby frog-hoppers (Fig. 51 a). Have you seen small brown or fawn insects about a quarter of an inch long, on the plants in the garden during the late summer or autumn, which, when touched, seem to jump away? These insects are called FROG-HOPPERS because they seem to hop, although, of course, they are really flying. Try to find some of these insects and look at them through a lens.

The frog-hopper has two large EYES, and two very tiny FEELERS. Like the aphis, it has a BEAK projecting from its mouth with which it pierces a plant and sucks up the sap (Fig. 51 c). Frog-hoppers have six legs and two pairs of WINGS. The first pair of wings are folded over and completely cover the abdomen, and are sloped like the roof of a house. The second pair of wings, with which it flies, are hidden underneath the first pair (Fig. 51 b).

During the summer, frog-hoppers lay their eggs on young shoots. A tiny green frog-hopper hatches out of the egg. Young frog-hoppers are similar to their parents but they have no wings (Fig. 51 *d*). The wings gradually develop as the frog-hoppers get older. The young frog-hopper sucks the sap of the plant with its beak. After the sap has been digested a sticky, watery fluid comes out of the end of the food canal or anus, which gradually covers the frog-hopper. Two glands at the end of its body give out a waxy substance, which

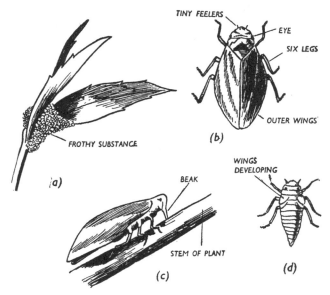

Fig. 51. *Frog-hopper:* (*a*) '*cuckoo spit*'; (*b*) *dorsal view of adult;* (*c*) *adult sucking sap from plant;* (*d*) *larva*

mixes with the watery fluid to make a soapy solution. The young frog-hopper then tips the end of its abdomen into the air, and with the help of two little projections at the end of its body, grasps a bubble of air and pulls it down into the soapy solution, so making it frothy. The frothy substance is called 'CUCKOO SPIT'.

Take one of the young frog-hoppers out of the 'cuckoo

spit' and look at it through a lens. You will see its EYES and its BEAK very clearly. Its body is SEGMENTED and it has six legs. If you look at a fairly large young frog-hopper

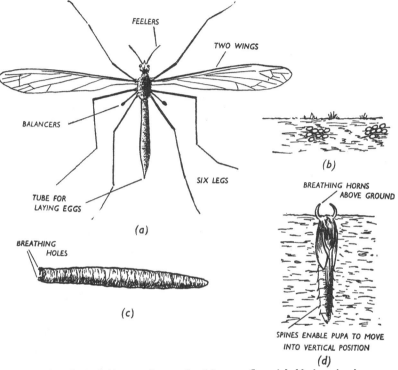

Fig. 52. *Life history of crane fly:* (*a*) *crane fly or 'daddy-long-legs';* (*b*) *eggs laid in the ground;* (*c*) *larva or 'leather jacket';* (*d*) *pupa*

you will see lumps on each side of the thorax (Fig. 51*d*). These are the wings, which develop gradually.

Caterpillars of butterflies and moths, aphides and frog-hoppers are all insects which do much harm in the garden. There are many more harmful insects to be found in the

91

garden but we cannot describe them here. Crane flies or daddy-long-legs whose larvae are called leatherjackets (Fig. 52) and click beetles whose larvae are called wireworms (Fig. 53) are very harmful and their life histories are illustrated in Figs. 52 and 53. The larvae of both insects attack

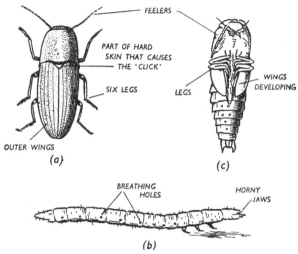

Fig. 53. *Life history of click beetle: (a) click beetle; (b) larva, called a 'wireworm', lives 4 years in soil, (c) pupa*

the roots of plants. They can be kept in school in a vivarium containing small plants, as the larva of the crane fly, and the larva and adult of the click beetle eat the roots of plants. Place the click beetle on its back in your hand. With a loud CLICK and a violent jerk it will jump on to its legs.

Lady-birds

Lady-birds are very small beetles that are familiar to everyone. There are several different kinds of lady-bird which vary not only in size but also in colour. The most common ones are reddish or yellow with black spots, or black with reddish or yellow spots.

Look carefully at a lady-bird and compare it with Fig. 54d. It has two compound EYES and two short FEELERS. Look at the head through a lens and you will see its MOUTH. It has

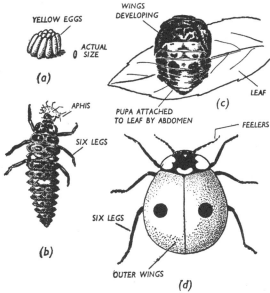

Fig. 54. *Life history of a lady-bird:* (*a*) *eggs;* (*b*) *larva;* (*c*) *pupa;* (*d*) *adult*

six LEGS and two pairs of WINGS. The first or top pair of wings cover the body and are coloured and horny. Carefully folded away beneath these wings there is another pair of filmy wings which are very large, and which are used for flying.

When you hold a lady-bird in your hand it gives out a nasty smelling liquid. This liquid protects the lady-bird from its enemies.

Lady-birds are very useful insects in the garden because they eat aphides.

Mother lady-birds lay clusters of 20 to 30 bright yellow, pointed eggs on the underside of leaves of plants that are infested with aphides (Fig. 54a). In a few days the EGGS hatch into peculiar looking, black LARVAE which have white or yellow spots on them (Fig. 54b). The larva has six LEGS and a fairly long, segmented body, but it has no wings. It walks along the stems of plants eating all the aphides that it can find. As the larva grows bigger it moults until it is fully grown. It stops eating and fastens the tip of its ABDOMEN to a leaf or to a stem. It then becomes shorter and fatter and changes into a PUPA (Fig. 54c). The pupa seems to be bent over on the leaf, and will move its head when it is touched. At first the pupa is bright orange in colour, but it gradually changes to dark brown and then to black. Look at a pupa through a lens and you will see the impression of its wings inside the pupal case. In about a fortnight the lady-bird emerges from the pupal case and at first it has no spots.

These insects are very easily kept in an insect box (Fig. 47), if you place shoots covered with aphides in a jar of water. You will see all the stages in the life history. Lady-birds hide during the winter amongst the dead leaves in the garden, and wake up during the first sunny days in March.

Hover-flies

Hover-flies are about half an inch long and can be seen hovering over flowers or darting through the air during the summer. Catch one of these flies and look at it carefully. They look something like bees because they are yellow and black in colour, and many people think that they can sting, but they are harmless. A hover-fly has only two WINGS whereas bees and wasps have four wings. It has two very large compound EYES, two small FEELERS (Fig. 55*a*) and six LEGS. Hover-flies suck the nectar and pollen out of the flowers.

The hover-fly finds a plant that is infested with aphides, and lays its cylindrical, pale yellow eggs amongst the aphides. In a few days the eggs hatch into dirty white, greenish or

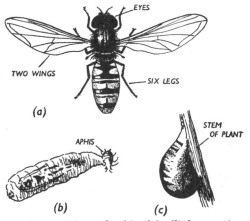

Fig. 55. *Life history of hover-fly: (a) adult; (b) larva eating aphis; (c) pupa attached to flower stem*

pinkish MAGGOTS or LARVAE which have no LEGS (Fig. 55*b*). Their bodies are flattened and are pointed at the head end. These maggots are very useful in the garden because they eat a very large number of aphides. When fully grown the maggot changes into a PUPA similar to that of a lady-bird only it does not 'stand up' at right angles to a leaf, but lies flat. After about ten days the case of the pupa splits and out crawls a hover-fly.

Hover-flies usually die at the beginning of the winter, but both maggots and pupae live throughout the winter hidden

95

amongst the dead leaves beneath plants. You may find them on the tightly packed leaves of a cabbage.

Larvae may be kept in an insect box if young shoots, infested with aphides, are kept fresh in a jar of water. You may see them change into pupae and hover-flies, but it is not easy to see or to find the eggs.

Ichneumon flies

Ichneumon flies are not commonly known to most people. You may, however, have seen small, white legless maggots wriggling out of the bodies of caterpillars. These are the LARVAE of one kind of ichneumon fly. Catch a number of

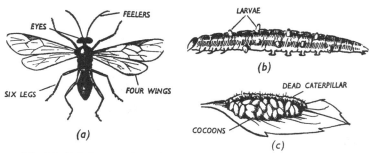

Fig. 56. *Life history of an ichneumon fly: (a) fly; (b) larvae emerging from caterpillar's body; (c) cocoons containing pupae*

Cabbage White butterfly caterpillars during August and September and keep them in your insect boxes, feeding them on cabbage leaves. When the caterpillars are big enough to change into pupae, you will see that many of them become very fat. They climb to a sheltered spot, or to the top of the insect box to pupate, but before they change into a pupa 20 or 30 small, white maggots about one-tenth of an inch long, eat their way out of the caterpillar's body (Fig. 56b). The caterpillar dies and shrivels up. Meanwhile, each of the little legless maggots spins a COCOON round itself and changes into a PUPA (Fig. 56c). You can see the maggots spinning their cocoons if you watch them closely. Many children think that these cocoons are the caterpillar's eggs, but caterpillars cannot lay eggs. If you keep some of these cocoons they will hatch the same year, or the following spring into small, black ichneumon flies. There are many different kinds of ichneumon flies which vary in colour and size.

Ichneumon flies lay their EGGS inside the bodies of caterpillars by means of a long, pointed tube. When the larvae hatch out of the eggs they live on the fluids that are inside the caterpillar's body. When they are fully grown they eat their way out of the caterpillar's body, and by so doing they kill it.

Look at an ichneumon fly through a lens and compare it with Fig. 56a. Its body is divided into three parts. It has two EYES, four WINGS and six LEGS. Some large ichneumon flies have a long abdomen which is pointed at each end forming a waist where it joins the thorax.

Social insects

Social insects are insects such as ants, bees and wasps that live together in colonies.

In this book we shall study small ants because they are easily found, and can be kept in school in a special ant box, called a FORMICARIUM (Fig. 57). Hive-bees can only be studied if you have a hive and all the equipment that is necessary for bee-keeping.

Ants

Look for ants between March and September in your garden, in fields, hedgerows, orchards or woods. When you have found a colony of small ants, dig out part of the colony with a trowel and put it into a covered jar. Make quite sure that you have one or two very large or QUEEN ants as well as the smaller WORKER ants. You will also find amongst the soil that you have dug up, some very tiny, white EGGS and some larger, white PUPAE which are commonly called 'ANT

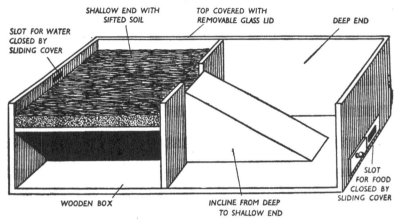

Fig. 57. *A formicarium. (The front of the box has been removed to show how the interior is constructed.) The shallow end must be kept dark by covering that end of the glass lid with brown paper or a piece of wood*

EGGS'. Place some very fine soil in the shallow part of the formicarium (Fig. 57) and then put the ants, eggs, larvae, pupae and soil that you have dug up into the deeper end of it. Cover over the shallow part to make it dark, but leave the deeper half uncovered. Ants do not like the light and will carry the eggs and young ones into the dark, shallow end. Place dead insects, or a little honey or jam, in the deeper end through the slot, for the ants to eat. Occasionally moisten the soil that is in the shallow end by running a little water from a pipette through the hole.

Under natural conditions ants make tunnels in the soil which have, here and there, larger holes or 'chambers'. The ant scrapes away the earth with its front legs and with its jaws.

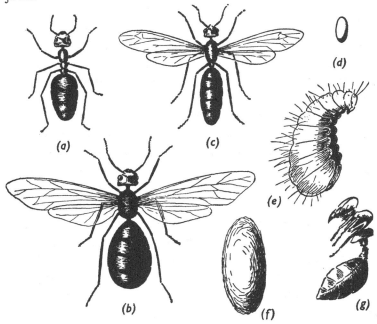

Fig. 58. *Ants:* (a) *worker;* (b) *queen;* (c) *male;* (d) *egg;* (e) *larva;* (f) *cocoon;* (g) *pupa.* [*These diagrams are not drawn to the same scale*]

Look carefully at a small worker ant and compare it with Fig. 58 a. Its body is divided into HEAD, THORAX and ABDOMEN. It has a pair of EYES and two large FEELERS. You may be able to see its biting JAWS. The thorax is smaller than the abdomen, and has six LEGS growing out of it. A queen ant is very similar to a worker ant but it is twice as big (Fig. 58 b). There may be several queens in one colony.

4-2

If ants find a dead insect they carry it back to their colony for food. It is an amazing sight to see many small ants carrying a dead wasp. Ants are very fond of sweet juices, and go in search of aphides in order to get the 'honeydew'. An ant will stroke the body of an aphis with its feelers, so making the aphis give out more 'honeydew' from its anus. Queen ants are fed by the workers. The workers swallow their food and later throw back some of it from their stomachs into their mouths. This food is given to the queen on the worker's tongue.

Only the queens lay eggs. The EGGS are about $\frac{1}{25}$th of an inch in size and are dropped by the queen anywhere in the nest. The workers pick up the eggs and carry them to NURSERIES where special ants act as nurses and look after the eggs. You should be able to see these special nursery chambers in your formicarium (Fig. 57). The eggs hatch into legless LARVAE (Fig. 58e) which are fed by the nurse ants on liquids from their stomachs which they throw back or regurgitate. The nurses lick the larvae to keep them clean, and put all the larvae that are the same size together in separate chambers. The nurses carry the young ants in their mouths. When the larvae are fully grown, the nurse ants cover them with soil. Inside the soil the larva spins a COCOON, inside which it changes into an ant. As soon as a cocoon is formed the worker ants remove the soil that was put over it. If you have one of these cocoons or 'ants eggs' as they are commonly called, remove the cocoon and you will see the PUPA inside it (Fig. 58f and g). When the pupa has changed into an ant, the workers cut up the cocoon along one side, and help out the pale-coloured ant. The workers then help her to unfold her legs and also feed her until her skin has become dark. Then she is able to look after herself.

During a hot day in July you may see hundreds of worker ants running about on top of the ground and with them you will see the large queen ants with wings, and also medium-sized ants with wings. The latter ants are the male or father ants (Fig. 58 c). All the winged queens and winged male ants fly up into the sky on a marriage flight. After the marriage flight the winged male ants die. A winged queen returns to the soil, either to her old colony or to a fresh place. She then pulls off her wings. If she returns to the old colony the workers will look after the eggs that she lays, but if she begins a new colony she has to look after the eggs and larvae until the first batch of worker ants have hatched out.

Ants spend the winter buried fairly deeply in the soil. Some of the larvae which hatch during the autumn do not change into ants until the following spring.

WASPS are very interesting insects to study, but as they cannot be kept under observation in school, they are not described here. The life history of a wasp is described in one of the books mentioned in Appendix C.

Hive-bees

Only a little can be said here of the hive-bees, but in many ways they are similar to ants. There are three types of bees in a hive (Fig. 59), just as there are three kinds of ants in the ant colony. There is only one QUEEN in a hive, and she is bigger than the other bees. The WORKER bees are the smallest bees, and the medium-sized bees, called DRONES, are the males. If you have a hive and bees at school, you will be able to look at these insects and compare them with Fig. 59.

The workers do all the work of the hive. When the bee is a certain age wax oozes out of her body and this wax is used to make the COMB of six-sided cells. Part of the comb is kept

101

as the 'nursery' and is called the BROOD CHAMBER, the remainder of the comb is a store-house for honey or pollen.

The queen lays one egg in each of the cells of the brood chamber. Each worker has a different job to do as she gets older. Her first job is to clean out the cells when a new

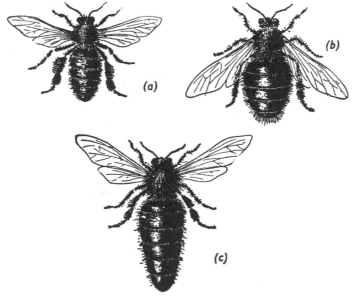

Fig. 59. *Hive-bees: (a) worker; (b) drone; (c) queen*

insect has emerged. Her next job is to get food from the stores and to feed the young larvae when they hatch (Fig. 60). Her third job is to take the nectar from the mouths of the old workers who have been out collecting the nectar and the pollen from the flowers. Inside her body the nectar is changed into honey which she regurgitates and squirts into the cells. By this time the wax is secreted from her abdomen, so that she is able to build new cells on the comb. She then becomes

102

a doorkeeper and either stands vigorously moving her wings to ventilate the hive, or else she stands on guard, smelling bees with her feelers, and driving away all strangers. Her

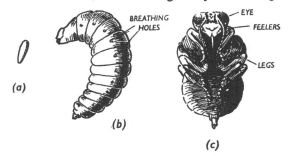

Fig. 60. *Life history of a hive-bee:* (a) *egg;* (b) *larva;* (c) *pupa*

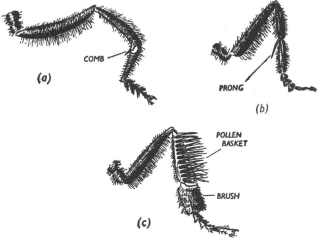

Fig. 61. *Legs of hive-bee:* (a) *front leg;* (b) *middle leg;* (c) *hind leg*

last job is to go out and to collect nectar and honey from the flowers.

Look at the LEGS of a worker bee and compare them with Fig. 61. The first leg has a comb which the bee uses to clean its head after it has been inside a flower. The second leg has a 'prong' which the bee uses to dig the pollen out of the pollen basket. The third leg has a pollen basket in which the bee stores the pollen that it has collected from the flowers, and also a brush which sweeps the pollen from the hairs of the body into the pollen basket of the opposite leg.

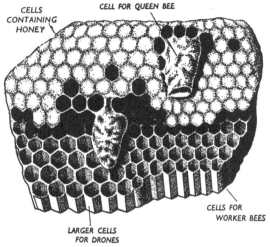

Fig. 62. *Piece of brood comb of the hive-bee*

When a hive becomes overcrowded, the workers build special royal cells (Fig. 62). The larvae inside these cells are fed on 'royal jelly' which makes them grow into queen bees. Meanwhile, in the brood chamber you will see some cells a little larger than the other cells. Out of these cells drone bees emerge. The old queen flies away from the hive, before the new queen emerges, followed by half of the workers. She settles down somewhere, and the workers all fly to this spot and form a solid mass of bees around her, which is called a SWARM. Some of the workers fly away as scouts to find a new home for the swarm. Meanwhile, a new queen hatches out in the hive and goes on a marriage flight with the drones. She then returns to the hive. The workers kill the remaining queen larvae.

In the autumn the workers kill the drones by stinging them, as they are lazy insects. The queen and the workers sleep throughout the winter, occasionally waking up to feed on the food that they have stored.

A bee's sting is barbed and when a bee stings the whole sting is usually pulled out of the bee's body, killing the bee. Consequently bees will not sting unless they have to. The sting of a wasp, on the other hand, is not barbed, and wasps can sting their prey as frequently as they please.

CHAPTER 5

ANIMALS OF THE COUNTRYSIDE

We have so far learned something about the animals that live in ponds, streams and gardens. We must now study the larger animals that we should see if we went for a walk in the country.

Reptiles

The only reptiles that live wild in the British Isles are grass snakes, vipers or adders, lizards and slow-worms. The last-named are not found in Northern Ireland or Eire. All these animals may be kept in school if you put them into a suitable vivarium (Fig. 63). They do not live happily in a very confined space.

The bodies of all these animals are covered with scales, and from time to time they shed their skins or SLOUGH, and a new skin grows in its place. (When insects lose their skins we say that they moult; see p. 32.) Fish, amphibians and reptiles are COLD-BLOODED animals. This does not mean that their blood is cold, but that the temperature of the body varies as the temperature of the surroundings rises or falls. We are WARM-BLOODED and the temperature of our bodies remains the same unless we are ill. All reptiles have NOSTRILS and breathe by means of LUNGS. Reptiles HIBERNATE or sleep during the winter, as the weather is too cold for them and they cannot get enough food.

106

Grass snakes and adders

All snakes are specially adapted to their mode of living. Try to find a snake in ponds, ditches, chalk hills or on sandy heaths. If you cannot find one you may be able to buy a grass snake from a livestock dealer. (See Appendix A.)

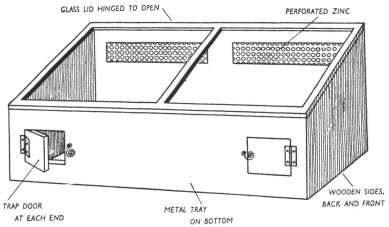

Fig. 63. *A vivarium for reptiles. Place soil and sand in the bottom, and plant grass and any small plants. Put large stones into the vivarium to make caves or hiding places for the reptiles. A small dish of water can be sunk in the soil to look like a pool*

Before looking for snakes you must be able to tell an adder from a grass snake, as the former has a poisonous bite (Fig. 65). A fully grown adder has a shorter and thicker body than has a grass snake. It may be two feet or two feet three inches long, whereas a grass snake may be three to four feet long. An ADDER has a flattened head which broadens behind the eyes, so that the head is distinct from the body. The eyes are coppery red in colour and the pupil is like a long slit. On the head it has a pair of dark bars which form a ∧ or an X, and there is a dark, wavy or zig-zag line down the

107

centre of the back. The general colour of the snake may be brown, olive or grey, but sometimes the ground colour is so dark that the darker markings are not easily seen.

The grass snake is grey, olive or brown in colour along its back, but it may be black and white or entirely black on the

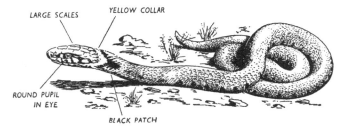

LARGE SCALES   YELLOW COLLAR

ROUND PUPIL
IN EYE

BLACK PATCH

Fig. 64. *Grass-snake*

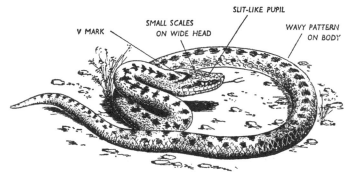

SLIT-LIKE PUPIL

V MARK   SMALL SCALES
ON WIDE HEAD

WAVY PATTERN
ON BODY

Fig. 65. *Adder or viper*

underside. It has a yellow or white collar behind its head, behind which are two patches of black. Its large eyes are golden or dark brown in colour and it has round pupils. The scales on a grass snake's head are much bigger than those on the head of the adder.

Snakes have no LIMBS. If you touch the ribs in your body you can feel that they are joined to the backbone on the

108

dorsal side, and to the breastbone or to another rib on the ventral side. Snakes, however, have RIBS throughout the whole length of the body, which are free on the ventral side, because there is not a BREASTBONE. The ribs act as legs, which do not project out of the body. Each rib is attached to a scale which serves as a sole.

If you can look at a live snake, watch its EYES very closely, and you will see that it does not blink. Its EYELIDS, which are transparent, are joined together and form an extra covering over the eye. When the snake sloughs, this extra skin is shed and a new one grows in its place. Snakes have EARS inside the head which are not very well developed, but they have no external ears.

A snake uses its TONGUE, which is forked at the end, for feeling its way. Although the tongue is continually moving in and out of the mouth, the jaws do not open, as the tongue protrudes through a gap in the front part of the upper jaw.

Snakes always swallow their prey whole and alive. Adders eat mice, shrews, voles, birds, lizards, frogs, newts and large slugs. Young adders eat insects and worms. Grass snakes can swim, so they may eat any amphibians or fish as well as mice and young birds. Young grass snakes eat worms, tadpoles, and young newts. If you keep snakes in school you need only feed them once a week, but you must give them their food alive. The body of a snake is specially constructed so that it can swallow animals that are fatter than it is itself. The TEETH point backwards to enable the Snake to get a good grip of its prey (Fig. 66). The adder has POISON GLANDS in the head from which small channels carry poison to the hollow FANGS (Fig. 66). The two halves of the lower jaw are not firmly fixed together in front as ours are, but they are joined by a substance which can stretch like elastic. Also the lower jaw is hinged with the skull in such a way that the jaws can gape widely. As the prey is swallowed the jaws widen, then the ribs, which are free, move upwards and outwards and allow the body to stretch. A python was once found that had swallowed a large live pig.

Grass snakes lay about 12 to 30 EGGS which are all connected in a string. Each egg has a tough, parchment-like

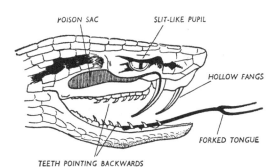

POISON SAC   SLIT-LIKE PUPIL

HOLLOW FANGS

FORKED TONGUE

TEETH POINTING BACKWARDS

Fig. 66. *Head of a poisonous snake*

shell. The mother snake places her eggs inside a heap of rotting plants or stable manure, which is always warm because heat is given off when organic matter decays. As soon as the eggs are laid they absorb moisture and swell up to one and a quarter inches in length. They hatch in six to ten weeks, and the baby snakes are six to eight inches long. Grass snakes often hiss, but they do not bite. The adder's eggs remain inside the mother's body until they hatch, so that the baby adders are born alive. Animals that lay eggs are said to be OVIPAROUS, whereas animals whose young are born alive are said to be VIVIPAROUS.

Snakes hibernate under the roots of trees or under piles of dead leaves.

110

Lizards

The lizards that are found in the British Isles are only about five inches long. Many children mistake newts for lizards, as both animals have similarly shaped bodies with long tails and four short legs. A lizard's body, however, is covered with SCALES, and a newt's skin is naked. You may find a lizard on a sunny, heath-clad hillside. Notice how

Fig. 67. *Common lizard*

quickly it runs. Do not pick up a lizard by its tail, because the tail will snap and the lizard will run away tail-less. A short tail may gradually grow from the stump.

You can keep lizards in a vivarium (Fig. 63) if you feed them regularly on flies, beetles, moths, caterpillars or spiders. These animals feed on small insects because they have a fixed mouth, like other animals, and their ribs are joined to the breastbone. The common lizard is brown or yellow-grey in colour with darker spots in longitudinal rows down the body. The tail is as long as the head and the body. The four limbs are almost equal in length and each one has five CLAWS. Lizards have upper and lower, movable eyelids, as we do, but in addition they have a third transparent eyelid, which

111

can be withdrawn to the inner corner of the eye. Lizards do not have external ears (Fig. 67).

Lizard's EGGS usually hatch immediately before they are laid, so that the young are born alive. There are about six to twelve young ones, which are dropped anywhere. They remain where they are for several days and are only about one inch long.

Fig. 68. *Slow-worm*

Slow-worm

If you find a slow-worm, you may at first think that it is a snake, but it is really a lizard that has no legs (Fig. 68). It is about 12 to 18 inches long. Slow-worms differ from snakes because, like the lizards, they have movable eyelids, fixed jaws, and only a few ribs, which are joined to a breastbone. They have teeth that point backwards and they eat spiders, small worms and slugs. They are grey in colour. If you stroke a slow-worm from head to tail it is extremely smooth. They usually slough three or four times a year. Slow-worms, like lizards, are always ready to part with their tails, consequently the end of their body may be ragged where the tail has broken off.

112

The slow-worm retains her eggs inside her body until they hatch. In August or September she produces six to twelve young ones each about two inches long. The young slow-worms are able to look after themselves straight away, and catch tiny insects and slugs for themselves. Lizards and slow-worms hibernate during the winter.

Birds

Birds live in almost all parts of the world. You may study birds in the town, in the country and in your own garden. Look carefully at any bird that you see and try to recognize it with the help of the books mentioned in Appendix C.

How birds are adapted for flight

Have you ever wondered why birds are able to fly? The reason, of course, is that their bodies are specially constructed to enable them to do so.

Bones

The body of a bird is light compared with that of a land animal of similar size. The reason for this is that the bones contain many AIR SPACES and so they are very light.

Wings

The front limbs are modified to form wings, which are moved by very powerful muscles. These muscles are joined to a special KEEL on the breastbone, which projects outwards from this bone (Fig. 69). The next time that you have a bird for dinner, look at its breastbone when the meat is removed, and you will see the keel. The wing is divided into the UPPER ARM, the FOREARM and the HAND, but there are only three

113

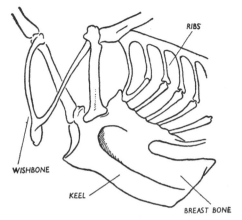

Fig. 69. *Breastbone of a fowl*

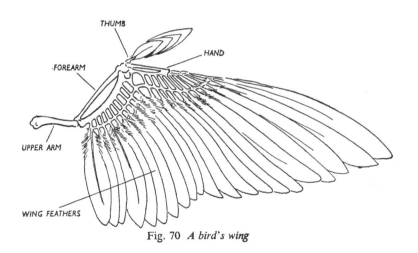

Fig. 70 *A bird's wing*

fingers, or DIGITS as they are called, the largest of which is the second (Fig. 70). These three parts of the wing are bent upon one another in the form of a Z when the wing is at rest. A fold of skin between the forearm and the body and another one between the forearm and the hand region give a larger surface to the wing. There are 23 large WING FEATHERS connected with the forearm and the hand as well as many covering feathers. As the bird is flying, the wing must move downwards with a solid surface, so that it can press on the air and so support the bird. As the wing comes upwards, however, it must not press on the air or the bird will be forced down again. The air must be able to pass through the feathers.

Tail

There are 12 long TAIL QUILLS which also help the bird to fly. These feathers are arranged horizontally in a semicircle on the tail and in addition there are many covering feathers.

Feathers

Look at a wing feather and compare it with Fig. 71 a. There is a central QUILL from which BARBS grow. On each side of the barbs there are BARBULES. If you look at a piece of the feather under a microscope you will see that the barbules, on the side of the barbs facing the tip of the feather, have HOOKS on them which face upwards (Fig. 71 b). These hooked barbules interlock with the barbules of the next barb, so that the feather knits solidly together and presses on the air as the wing goes down. As the wing comes up it does not press on the air, as the air can pass through the barbules. The feathers which cover the body do not have hooks on all the barbules. Young birds are covered with DOWN FEATHERS (Fig. 71 c).

115

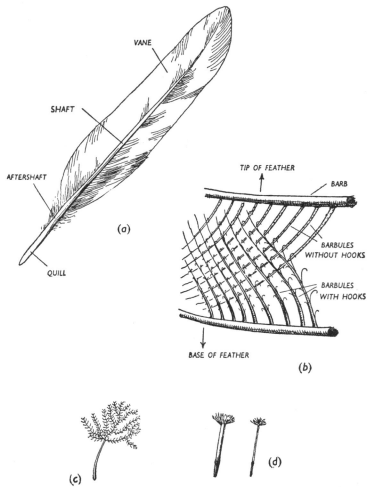

VANE

SHAFT

AFTERSHAFT

QUILL

(a)

TIP OF FEATHER

BARB

BARBULES
WITHOUT HOOKS

BARBULES
WITH HOOKS

BASE OF FEATHER

(b)

(c)

(d)

Fig. 71. *Bird's feathers.* (*a*) *Whole feather;* (*b*) *part of feather enlarged;*
(*c*) *down feather;* (*d*) *pin feathers*

116

PIN FEATHERS (Fig. 71*d*) are sometimes found amongst the feathers that cover the body. Feathers sometimes fall off the bird's body and new feathers grow in their place. This is called MOULTING. During the winter the birds grow more feathers to keep them warm.

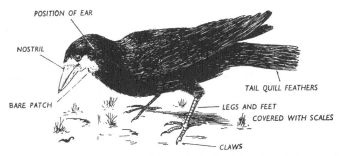

Fig. 72. *A rook. The bare patch behind the beak distinguishes it from a crow*

Breathing

Between the feathers and the skin there is a layer of air which keeps the bird warm. Birds need a lot of energy when they are flying, and to get this energy their temperature is higher than ours, and in addition they need a very good supply of oxygen. As a bird breathes in, air not only fills the LUNGS, but also passes through the lungs into nine AIR SACS, some of which lie partly in the bones. Most of this air is stored in the sacs, but some of it passes out of the body as the bird breathes out. This means that the blood in the lungs can get oxygen when the bird breathes in, and also it can get oxygen from the stored air as the bird breathes out.

Eyes

Look at a bird's head. You will see its NOSTRILS, through which it breathes, and its EYES (Fig. 72). Birds' eyes are very large and have upper and lower eyelids as we have, and in addition they have a third eyelid or NICTITATING MEMBRANE which stretches across the eye from the inner corner. When birds are flying quickly, the wind presses very strongly on their eyes, so the eyes are strengthened with special BONY PLATES.

Ears

The ear of a bird does not have an external ear lobe projecting from the head. There is a small opening behind the eye, which is protected by a circular patch of small feathers, the EAR COVERTS. The opening leads into a small tube, at the bottom of which is the EAR DRUM.

Beaks

The shape of a beak will tell you what a bird eats (Fig. 73). Birds with tiny beaks, like those of the sparrows, eat seeds. Thrushes and blackbirds have longer beaks with which they catch worms and insects. Herons and kingfishers, which you may see along river banks, have long beaks with which they catch fish. Hawks, kestrels, owls and eagles have very sharp, curved beaks for tearing flesh. The beak of a duck is flattened horizontally to enable it to strain small creatures out of the mud. Although parrots and budgerigars are not wild in this country, we often keep them as pets. Look at their beaks and you will see that they are strong and hooked to enable them to crack seeds or nuts. Birds have no TEETH. When the food is swallowed it goes into the CROP, which can easily be seen when it is full, bulging out to the right at the base of the neck. Here the food is mixed with digestive liquids. A little at a time the food passes into the thick-walled GIZZARD, which is usually full of stones which the bird has

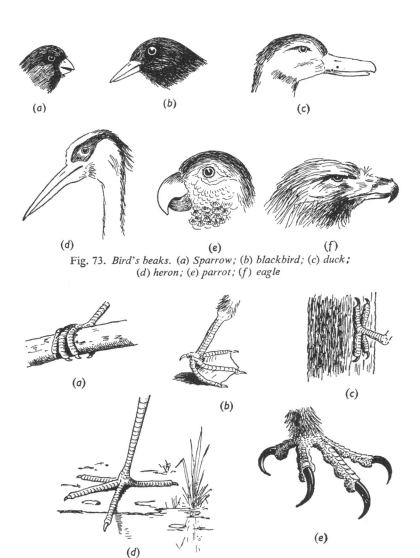

Fig. 73. *Bird's beaks.* (*a*) *Sparrow;* (*b*) *blackbird;* (*c*) *duck;* (*d*) *heron;* (*e*) *parrot;* (*f*) *eagle*

Fig. 74. *Bird's feet.* (*a*) *Pigeon;* (*b*) *duck;* (*c*) *woodpecker;* (*d*) *heron;* (*e*) *eagle*

swallowed. As the walls of the gizzard move, the food is ground up by the stones. Birds that eat flesh do not have a hard-walled gizzard.

Feet

The feet of all birds are covered with SCALES (Fig. 74). If you look at a bird's feet you will be able to tell roughly where it lives. Perching birds, such as the sparrow and the pigeon, have three toes pointing forwards and one toe pointing backwards. Climbing birds, like the woodpecker, have two toes pointing forwards and two toes pointing backwards. Swimming birds, like the duck, have webbed feet, but the moorhen swims in the water although its feet are not webbed. The heron walks in shallow water, and so it has long, spreading toes which prevent it from sinking into the mud. Hawks, kestrels, owls and eagles have sharp claws or talons with which they hold their prey.

Nests and eggs

Birds are very good parents and take care of their young ones. Father birds often have brightly-coloured feathers, whilst mother birds have dull-coloured feathers which cannot be seen very easily.

Domesticated birds, such as fowl, ducks, etc., lay eggs at any time of the year, but wild birds lay their eggs during the spring. All birds, except the cuckoo, make a nest in which they put their eggs. See how many nests you can find, but do not touch them during the nesting season. Each kind of bird has its own way of making a nest, which is built in a special place. Thrushes and blackbirds build their nests in bushes. Both nests are made of mud and grass, but the blackbird puts the mud on the outside of its nest, whilst a thrush lines its nest with mud which it smoothes with its breast. Rooks make coarse nests of twigs which are built high up in the tree-tops. Robins make soft little nests of moss, horses' hair and feathers, which are hidden at the bottom of a hedgerow.

Birds that build closed nests lay white eggs, whereas birds that lay their eggs in the open have coloured eggs, which match the surroundings so that they cannot be seen. Eggs

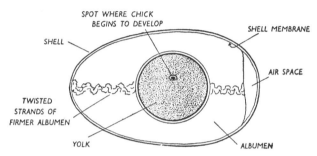

Fig. 75. *A bird's egg with top of shell removed*

that are in a deep nest are round in shape, but eggs that are laid in flat nests are pointed at one end so that they cannot roll out of the nest.

Fig. 75 shows you a picture of a hen's egg with half of the shell removed. The YOLK is the food on which the baby bird feeds as it develops. Look carefully at the yolk and you will see a WHITE SPOT on the top of the yolk. It is from this white spot that the young chicken begins to develop when the hen sits on the egg. The yolk is protected by the ALBUMEN or white of the egg, the SHELL MEMBRANE and the SHELL. At the blunt end of the egg, the shell membrane divides into two, and encloses an AIR SPACE. The two twisted strands of firmer albumen at opposite sides of the yolk enable the yolk to twist round so that the white spot is always on top.

A bird lays her eggs one at a time, but she does not sit on them to keep them warm until she has finished laying. The first eggs are usually cold before the last eggs are laid, but this does not harm them, as they do not begin to develop until the mother bird sits on them. When development has started the eggs must not get cold or they will die, so mother bird only leaves her eggs for a short time whilst she has a short flight to stretch her wings and to catch food. Father bird may also sit on the eggs, or he may fetch food for mother bird.

When the baby birds hatch, they have very wide mouths, and they are fed by their parents. The parents swallow the food, which is made soft in the crop. They then bring the softened food into their mouths again, and give it to their babies. Young domesticated birds can run about and feed themselves shortly after they have hatched out of the egg. The parents teach their babies to fly.

Migration

Insect-eating birds, such as swallows, swifts, martins and cuckoos, migrate in the autumn when the length of daylight is shorter. Nearly all the English birds that migrate fly to Africa, where they remain until the following spring when they return to breed. Some birds come to England, from colder countries, in the autumn.

Bird study

The study of birds is most interesting. Choose a few birds that you can see every day and watch them. You will soon learn to recognize them by their shape and colour and by their song. Each species of bird has its own peculiar habits, which you will notice when you watch them flying, walking or hopping, alighting on a branch, or feeding. Try to discover when and where they nest, the materials used in building the nest, and the number and colour of the eggs. Do not touch the nest or the eggs or the bird will leave them.

Birds as pets

Do not keep birds in captivity unless you can give them a large enough aviary for them to fly about. They should also be able to breed in captivity. If you have sufficient space in your school garden or playground you could keep poultry. Advice on the housing, feeding and breeding of poultry may be obtained from members of a local poultry club or from accredited breeders, or you may buy books and pamphlets which would give you information on the making of poultry houses and runs, the varieties of birds to breed and how to feed them (see Appendix C).

Mammals

Mammals are animals that are covered with HAIR. They all have four LIMBS, and many of them have TAILS. Their young ones are always born ALIVE and they are fed with MILK which is secreted from special GLANDS in the mother's body. There are many mammals, both wild and domesticated, living in this country. In addition, many foreign mammals may be seen in zoological gardens. Make a list of all the mammals that live in this country.

Hoofed mammals

Cows, sheep, pigs, horses, goats and deer are often seen in the country. If you look at the feet of these animals you will see that they all walk on HOOFS, which are really horny coverings over their toes. A horse has a solid hoof which covers the one toe on which it walks. It is a very fast runner, and, when wild, always runs on hard ground. The solid hoof

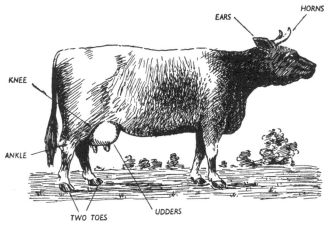

Fig. 76. *A hoofed mammal—a cow*

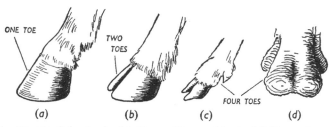

Fig. 77. *Hoofed animals.* (*a*) *Horse;* (*b*) *cow;* (*c*) *pig;* (*d*) *hippopotamus*

of a horse would stick in wet ground, and the horse would have great difficulty in pulling its foot out of the mud, as the solid hoof would act as a sucker. Cows and pigs walk on two toes, so they have a split or cloven hoof and not a solid one (Fig. 77). This enables them to walk on muddy ground.

If you look at a pig's foot you will see that in addition to the two toes on which it walks it also has two toes that project backwards (Fig. 77). Some large foreign hoofed mammals, such as the hippopotamus, walk on four toes, and the rhinoceros walks on three toes.

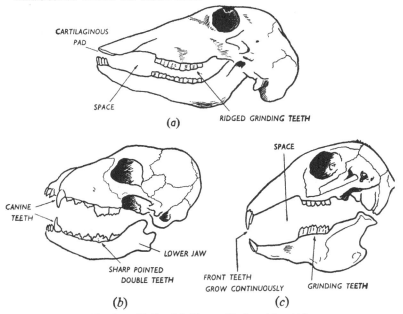

CARTILAGINOUS PAD

SPACE

RIDGED GRINDING TEETH

(a)

SPACE

CANINE TEETH

LOWER JAW

SHARP POINTED DOUBLE TEETH

FRONT TEETH GROW CONTINUOUSLY

GRINDING TEETH

(b)   (c)

Fig. 78. *Skulls.* (a) *Sheep;* (b) *dog;* (c) *rabbit*

You can also recognize a hoofed mammal by its teeth, as there is always a space between its front teeth and its back double teeth (Fig. 78 a). The double or GRINDING TEETH are squarish on top with small ridges which are used for grinding up the hay, grass or leaves on which these animals feed. A few of the animals in this group 'chew the cud'. Do you know what this means? When a cow is feeding in a meadow it bites and swallows its food without chewing

125

it. This food passes into the first part of the stomach (Fig. 79). When the cow has eaten enough food it lies down. Small balls of food pass into the mouth from the stomach. The cow chews this food thoroughly, and then swallows it. The chewed food passes into the second part of the stomach, where it is digested.

Many hoofed mammals go about in herds for protection. Some of them are very quick runners and can run away from their enemies, whilst others have HORNS or ANTLERS that may

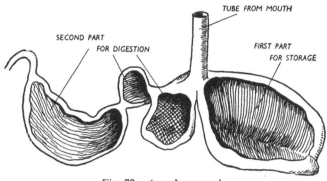

Fig. 79. *A cow's stomach*

be used for defence. At the base of the cow's horn there is a solid core of bone, which is covered with horn. These horns continue to grow throughout the animal's life. Deer, on the other hand, have antlers which are branched (Fig. 80). All reindeers have antlers, but of the deer that live in this country only the fathers or STAGS have them. Antlers are covered with a furry substance called VELVET. Every year the velvet peels off and then the antlers break off. New antlers grow during the spring and the summer. One stag lives with several mother deer or HINDS, and he protects them with his huge antlers.

126

In the spring each hind finds a protected place in the forest where her solitary baby is born. The baby animals belonging to this group are very well developed when they are born. A COLT or baby horse can run almost as soon as it is born,

Fig. 80. *A fallow deer*

as it is only safe amongst the herd. Nearly all hoofed mammals have only one young one at a time, but animals like the pig, that do not live in such open country, have many young ones which are not well developed when they are born and so they need more protection.

Beasts of prey

Cats and dogs belong to this group of mammals as well as the wild foxes, wolves, badgers, stoats, weasels, otters, etc. Look at the teeth and at the feet of one of these animals and compare them with those of the hoofed mammals. You will see that there is not a space between the front and the back teeth; instead, beasts of prey have four very long, pointed CANINE teeth, which, with the remainder of the teeth that are all pointed, tear the flesh of the animals that they eat (Fig. 78 b). Beasts of prey have five claws on each foot, which may be used for tearing flesh, for fighting their enemies or for climbing.

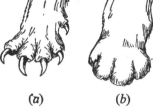

(a)   (b)

Fig. 81. *Cat's feet.* (a) *Showing claws;* (b) *claws hidden*

Cat

All the beasts of prey that are similar to a cat have claws that can be withdrawn (Fig. 81). This enables the animal to walk silently when it is stalking its prey. The claws are also kept sharp so that the animal may climb more easily. Look at a cat when it is sitting in the light and you will see that the pupil of each eye is like a vertical slit.

Dog

Dogs and their relations the foxes and wolves, cannot withdraw their claws, so they become blunt as the animal runs about. If a house dog is not allowed to run about very much its claws will grow too long and they will have to be cut. When wild these animals (except the fox) hunt in packs, so there is no need for them to creep silently up to their prey. In addition these animals do not climb, so they have no need for sharp claws. Dogs and cats have very large families. The babies are very tiny when they are born, and cannot even open their eyes until they are a fortnight old, so the parents take care of them.

Fig. 82. *A fox*

Fox

A fox is about two feet long and fourteen inches high at the shoulder. Its fur is russet brown on top and white under its body (Fig. 82). It has a long, bushy tail that is tipped with white and is called the BRUSH. Unlike the other members of the dog family, the fox lives alone in a home which is called its EARTH, which is usually a hole previously made by a badger or by a rabbit. Foxes come out at night to catch their prey, which consists of rabbits, hares, pheasants, partridges, hedgehogs, squirrels, voles or even frogs. In addition a fox will kill poultry and even tiny lambs. In April the mother fox or VIXEN has about four blind babies which can see when they are ten days old. Mother is able to take out her young ones when they are a month old, and she looks after them until the autumn.

Stoats, weasels and polecats

Stoats, weasels and polecats have long, thin bodies which enable them to run down holes after rabbits. They have very strong, sharp teeth with which they bite through the brain

Fig. 83. *A weasel*

case of their victim. The STOAT or ermine is the largest animal of the three and can be recognized by the black tip at the end of its tail. When winter comes, with a sudden fall of temperature, a stoat's brown fur changes to white. It hunts for fish in shallow streams, and for rabbits, and it destroys poultry and moles. It may hunt during the daytime or only at night. In April the mother stoat makes a nursery for her four or five babies in a hole in a bank or in a decayed tree.

The weasel is smaller than the stoat, but it is the same colour. It is only nine to ten inches long, and two inches of this is its tail. It has a narrow head. In spite of its size the weasel will attack animals that are much larger than itself. It eats rats, mice, voles, moles, frogs or birds. It can swim to catch the water voles, or with its sharp claws it can climb trees to get birds' eggs. Weasels will eat chickens. In this country the weasel's fur does not change colour in the winter, but weasels that live in colder climates turn white during the winter. The weasel has five or six young ones, which she keeps in a hole in a bank or in a tree.

A ferret is a domesticated polecat. It is two feet long with a tail that is seven inches long. It is extremely destructive, but fortunately there are very few wild ones in this country. Its fur is dark brown on its back, and black on the underside. Tame ferrets may be white or light brown in colour. The polecat usually lives in woods that are near to a farm where it can enter a henhouse. It will kill all the poultry although it may only want to eat one bird. Mother polecat makes a nest of grass in a hole and the five or six young ones are born during April or May.

Fig. 84. *A badger*

Badger

Badgers are very seldom seen, as they come out only at night and spend the whole of the day in their SET in the earth. The set many be ten feet underground and from it several passages lead to the openings above ground. In front of one of the openings the badger puts all the soil that he has excavated. A badger is two to three feet long and its shoulders are about one foot high. It is grey in colour with a black and white striped head. Its ears are short and it has a tail that is about eight inches long (Fig. 84). A badger will eat any small animals that it can find, and digs pits into which it puts all offensive waste matter. In the spring or the summer the mother badger makes a nest of moss and grass for her four or five babies, which are born blind and helpless. Badgers sleep during the winter in special deep chambers, the passages to which are blocked to keep out unwelcome visitors. If the weather is warm, badgers wake up and go in search of food.

Gnawing mammals

To this group of mammals belong all the animals that gnaw their food. They are all fairly small animals with very sharp CLAWS, and, like the hoofed mammals, they have a space between their front and their back teeth (Fig. 78 c). These animals are vegetarians, but rats and mice will eat anything. Gnawing mammals usually have very large families. The babies are kept in a nest because they are very immature when they are born. With the exception of the dormouse and the squirrel these animals do not hibernate, as they can obtain food during the winter. As in all mammals, their fur is much thicker during the winter than it is during the summer.

Rabbits

Rabbits are perhaps the best known gnawing mammals, and they can be kept very easily in hutches. If you do not know anything about the keeping of rabbits there are many books and some pamphlets (issued by the Ministry of Agriculture and Fisheries) that you can buy which will give you all the information that you require on the building of hutches, the feeding and breeding of rabbits and also what you must do if your rabbits are ill.

Rabbits' front teeth never stop growing, so they have to nibble hard things to wear away the teeth. Rabbits have very long ears to enable them to hear well (Fig. 85). This is very necessary because, although they are partially protected

Fig. 85. *Rabbits*

by their colour, their only method of defence is to run into their BURROWS when they hear a sound. The burrows are made in light, dry soil. The rabbit digs out the soil with its forepaws, and kicks the soil away with its hind feet. The back legs are much bigger and stronger than the front legs, because, as you know, the rabbit hops about. Unlike the beasts of prey, they do not have pads on their feet; instead the underneath part of the foot is covered with hair which gives a firm grip on hard rock or on slippery snow. Rabbits have very short tails, which are white underneath. When a rabbit hops

133

the other rabbits can see the white tail. Rabbits warn one another of danger by stamping their back feet on the ground. Mother rabbit has several families of three to eight young ones during the spring, summer and autumn. She makes a nest for them of dried grass and lines it with fur which she pulls off the underside of her body. This nest is made in a special burrow or STOP which is not used by other rabbits. If you keep tame rabbits you must not keep the father rabbit or BUCK with the mother rabbit or DOE when she has a family, as he may kill the babies. Young rabbits can neither see nor hear until they are ten or eleven days old. They are able to look after themselves when they are a month old. Rabbits will eat many plants and they do a lot of damage to crops.

Hares

Hares are very similar to rabbits but they have larger bodies and longer legs. They do not make burrows; instead they make a nest or FORM by flattening the grass amongst gorse or briar bushes, or even on open ground. Mother hare rears three or four babies in her nest. As the young ones are not hidden in a hole, they can see when they are born and they are covered with fur. Each baby makes its own little form near to that of its mother. It can live alone when it is one month old. Hares often damage the farmer's crops and young trees.

Squirrels

Squirrels live in trees, and have extremely long and sharp claws which enable them to climb. The English squirrel has reddish-brown fur with white fur on the underside of the body. The bushy tail is as long as the body and the squirrel curls its tail round its body to keep it warm when it is asleep.

It also helps the squirrel to balance. Squirrels have large tufted ears that are always erect, and large eyes. Squirrels eat seeds of cones, beech nuts and hazel nuts and even cherries. They hold their food in their short forepaws, which have only four clawed fingers and a very tiny thumb. The hind limbs, which are much longer than the fore limbs, have five toes. Squirrels make nests or DREYS of twigs, bark, moss

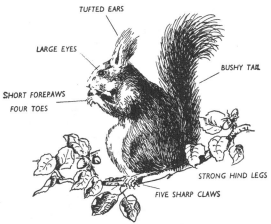

TUFTED EARS

LARGE EYES

BUSHY TAIL

SHORT FOREPAWS
FOUR TOES

STRONG HIND LEGS

FIVE SHARP CLAWS

Fig. 86. *A squirrel*

and leaves either in the branches or in the hollow trunk of a tree. Sometimes squirrels use old nests belonging to rooks or magpies. Mother squirrel may have two families in one year, one in the spring and one in the autumn. The mother either makes her nest for the babies in a hollow tree, or she makes a huge ball-shaped nest which has a side entrance. Here she has four or five blind babies, which remain with their parents until they are fully grown. Squirrels have very long sleeps during the winter, but they wake up occasionally to eat some of the nuts that they stored during the autumn.

135

Mice

Mice can quite easily be kept in school. It is wise to have an extra cage, so that the cages can be scrubbed out frequently. You must have a dark nesting box and plenty of room for the mice to run about. Many mice love to exercise their legs by running in a wheel (Fig. 87). Mice usually live quite happily

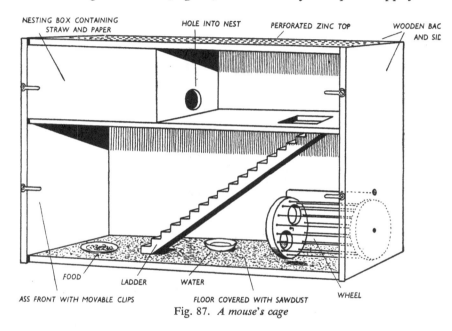

NESTING BOX CONTAINING STRAW AND PAPER   HOLE INTO NEST   PERFORATED ZINC TOP   WOODEN BAC AND SIC

FOOD   LADDER   WATER

ASS FRONT WITH MOVABLE CLIPS   FLOOR COVERED WITH SAWDUST   WHEEL

Fig. 87. *A mouse's cage*

together as a family, but some father mice kill the young ones. These animals have fairly large ears and eyes, and sharp claws with which they can climb. They have naked tails. Mice will eat any food that we eat, and they will nibble paper, straw, material, etc., which they put into their nests. If you watch a mouse eating you will see that it holds its food in its forepaws as the squirrel does. Mice are very clean animals

136

and are frequently washing themselves. They are very harmful to us because they run about anywhere and pick up germs on their feet and fur, then they may run on our food. Mice breed very quickly. They may have a family of six to twelve young ones five or six times a year. The babies are naked when they are born, with eyes and ears closed. The mother mouse makes a warm nest of dried grass, paper, material, etc., which she bites into small pieces. Young mice can breed when they are about six weeks old. It is wise to separate father mouse from mother mouse when she is going to have a family.

If you wish to keep mammals in a classroom, it is advisable to keep hamsters, if you have a suitable cage. These animals do not smell unpleasantly, as do mice. They are friendly little animals, and will make nests of paper, straw, etc. They will eat any food that we eat, especially oats, dried fruit, apples, nuts, biscuits and some garden leaves such as dandelion leaves. They quickly put the food into pouches in their cheeks, and take it to a hiding place near their nest. You can give them enough food to last for several days.

Rats

Rats are very similar to mice, but they are much bigger (Fig. 88). They are not fully grown until they are about six months old, and, like mice, have five or six families a year. They are extremely destructive to food and destroy anything that comes their way. They also spread diseases.

Insect-eaters

There are three well-known insect-eaters that can be seen in the country, the hedgehog, the mole and the shrew. All these animals have pointed snouts, very small teeth all round their mouth, and five claws on each foot.

Fig. 88. *A rat*

SPINES

SNOUT

CLAWS

SOFT HAIR

Fig. 89. *A hedgehog*

Hedgehog

The hedgehog is perhaps better known than the other two animals. Hedgehogs can be kept in a walled garden, or in a piece of garden fenced off from the remainder, or they may be kept in a very large hutch which has a darkened sleeping end. Even in the summer, hedgehogs like to sleep in a ball of dried leaves, grass or straw, so you must put straw into their hutch. They are very good climbers, so make sure that your hedgehogs cannot escape; they can also squeeze through

138

very narrow spaces. They will eat any small insects, snails, slugs, worms or even mice, rats, frogs, lizards, snakes and even bird's eggs. In captivity you may also feed them on bread and milk. Hedgehogs sleep during the daytime, and frequently snore very loudly. They come out at dusk to catch their prey. If you look at a hedgehog you will see its fairly large ears, its bright eyes, its pointed snout and its four little feet with sharp claws (Fig. 89). As you probably know, the whole of the body except the underside is covered with prickles, so the hedgehog can roll itself up into a prickly ball to protect itself from its enemies. During the summer, mother hedgehog makes a cosy nest of dried grass and leaves in a sheltered spot or in a hole and there she has a family of four to seven babies, which are blind and are completely naked. Small soft spines gradually grow which stiffen as the hedge-hogs get older. Hedgehogs will breed in captivity, but it is not wise to keep father hedgehog with the babies. Hedgehogs hibernate during the winter as they are not able to get food. Before hibernating they eat plenty of food, some of which is stored as fat in their bodies. They then make very warm nests in which they sleep.

Moles

Moles are not easily seen or caught, as they live almost entirely underground. They are covered with blackish-grey, short fur that can be stroked in any direction. As it lives in the soil it has very tiny little eyes, and no external ears. It has a long, flexible snout, with which it can burrow into earth that has been loosened with its large hands (Fig. 90 a). A mole's hand is very peculiar; it looks like our hand, only it is made entirely of horn, and the thumb cannot move (Fig. 90 b). The mounds of earth thrown up by the mole at frequent intervals as it burrows through the earth are commonly called MOLE-HILLS or mole-heaves, and are made to get rid of the soil when it is making a new burrow. The mole-hill or home where the mole lives is about one foot high and three feet broad and is usually hidden under a bush or hedgerow. Mother and

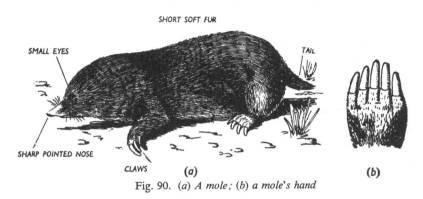

Fig. 90. (*a*) *A mole;* (*b*) *a mole's hand*

father moles live apart. The mother makes a nest of leaves or of grass in her mole-hill, and there, in April or May, she has a family of two to seven young ones, which are blind, naked and pink when born. Their eyes do not open for three weeks. Moles do not hibernate, as they eat any small animals that live in the soil.

Shrews

Shrews at first look like small mice, but it you look more closely you will see that they have a long, pointed snout. They eat tiny animals. They do not breed as quickly as mice do, as they only have about two families of two to eight young ones every year (Fig. 91).

140

POINTED SNOUT

NAKED TAIL

Fig. 91. *A water shrew*

Bats

There are several kinds of bat in this country which are alike in many ways. They can fly, and their bodies are so well adapted to flight that they can no longer move on land. The bat's body is similar to that of a mouse, only its face is flatter. Its WINGS are made of skin which stretches between its very long fingers, its feet and its tail (Fig. 92*a*). The THUMB forms a hook, with which the bat can cling to any rough surface. Bats, however, sleep upside-down, with their wings folded across their body (Fig. 92*b*). They hang on to a ledge with their feet. They come out at dusk to catch their food, which consists of night-flying insects. Only one baby is born at a time. It is blind, but it has a little fur on it. The young bat clings to its mother's fur for the first fortnight of its life, and then the mother can leave it behind when she flies around. Young bats cannot leave their mothers until they are two months old. Bats cannot see very well, but they can feel where they are flying by their sensitive whiskers and the extremely sensitive skin on their faces, ears and wings. Bats hibernate during the winter, as they cannot obtain food.

141

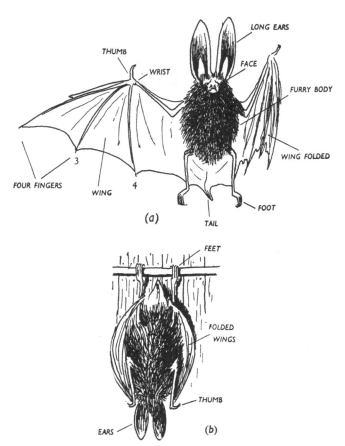

Fig. 92. (a) A long-eared bat; (b) a bat asleep

142

CHAPTER 6

FLOWERS AND SEEDS

Most of the plants that you know have ROOTS, STEMS, LEAVES and FLOWERS, each part having a different use for the plant.

The flower

The flower is a very important part of a plant, because in the flower, seeds are produced which will grow into new plants. Almost all flowers have the same parts, though they may differ in colour, number and shape. In this book we shall describe a wallflower, because there are so many flowers like it, some of which are open at all times of the year (e.g. Siberian wallflower, aubretia, cabbage, lady's smock, shepherd's purse, charlock, mustard, yellow rocket, bittercress, horse-radish). When you are looking at the wallflower, look also at several other flowers and compare them with it.

The wallflower

If you look at a wallflower (Fig. 93) you will see on the outside, leaf-like parts called SEPALS which together form the CALYX. The sepals protect the flower when it is a bud. In some flowers, such as the Deadnettle (Fig. 101), the sepals may be joined together.

Inside the sepals and alternating with them you will see four coloured PETALS which are brightly coloured to attract insects. These together form the COROLLA. In some flowers the sepals and petals are alike and together form the PERIANTH.

Pull aside the petals and the sepals and you will see six STAMENS. Each consists of a STALK or FILAMENT with a swelling at its tip called a POLLEN BOX or ANTHER. Each anther, when it is ripe, is divided into four compartments or SACS. Each sac contains a yellow powder which consists of tiny POLLEN GRAINS. The anther bursts when it is ripe, and the pollen grains are set free.

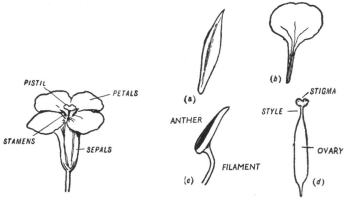

Fig. 93. *Wallflower*

Fig. 94. *Parts of a wallflower. (a) Sepal;*
(b) petal; (c) stamen; (d) pistil

In the centre of the flower you will see a PISTIL. Looking at the top of the flower, you will see a green knob in the middle of the stamens. This is the tip of the pistil and is called the STIGMA. Pull aside the sepals, petals and stamens and you will see that a short stalk or STYLE joins the stigma to the SEED BOX or OVARY where the young seeds or OVULES grow. Some flowers, such as the buttercup, have many pistils in the middle of the flower. All parts of the flower grow from the end of the flower stalk which is called the RECEPTACLE.

144

Pollination and fertilization

The ovules cannot grow into new plants unless some pollen reaches them. The pollen grains cannot actually reach the ovules themselves, as they are enclosed in the ovary. The pollen grains must first reach the stigma. They either fall on to the pistil, or they are carried to the pistil by wind or by insects. When the pollen grains have reached the stigma we say that the flower has been POLLINATED. The stigma secretes a sugary liquid which makes it sticky. In this sugary liquid

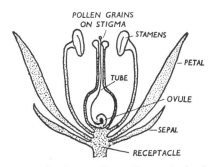

Fig. 95. *Section of a flower. Note pollen grain fertilizing an ovule*

the pollen grains begin to grow, and from each pollen grain a tube grows down into the ovary. (If you put some pollen grains into some sugar solution you will be able to watch these tubes growing.) One tube grows right into each ovule, where the nucleus of the pollen grain joins with the nucleus of the ovule. This is called FERTILIZATION. Flowers can only be fertilized by pollen from the same kind of flower.

The ovules grow into much stronger plants if they have been fertilized with the pollen from another plant. This is called CROSS-POLLINATION. Seeds that are pollinated with pollen from the same plant are said to be SELF-POLLINATED. Many

145

flowers are so made that they are not able to pollinate them-selves. Cross-pollinated flowers are pollinated either by wind or by insects.

Pollination by wind

Wind-pollinated flowers have no petals or scent to attract the insects. They have plenty of light pollen which is smooth on the outside and which can easily be blown about by the wind, and long, sticky or feathery stigmas which can catch the pollen.

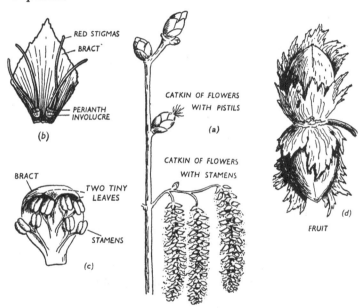

Fig. 96. *Hazel.* (a) *Twig with flowers;* (b) *pistillate flower;* (c) *staminate flower;* (d) *fruits*

The hazel

One of the first flowers to open in the spring is the hazel. Hazel flowers are not at all like the flowers of the wallflower.

146

Before the leaves appear, you will see long yellow CATKINS which wave about in the wind (Fig. 96). These catkins are made up of many very tiny flowers each consisting only of a BRACT, to which two tiny leaves are fused, and four stamens. The filament of each anther is split into two so that it looks as though there are eight stamens. When the catkins are shaken they give out a cloud of yellow pollen. Look closely at the green buds on the same twig and you will see that some of them have red threads (which are pistils) coming out of their tips. These buds contain the flowers with the pistils. The flowers are arranged in pairs, inside the bud, and each pair is protected by a bract (Fig. 96). Each flower consists only of a pistil which has two long red stigmas which lead into a seed box which contains two young seeds. This is surrounded by a minute lobed perianth around which there is a collection or INVOLUCRE of bracts. When the wind blows, pollen from the catkins is blown about, and some of it will fall on the long red stigmas. The fertilized seeds will grow into hazel nuts. Hazel is usually pollinated with pollen from the same tree, but it could be cross-pollinated if another hazel tree was growing next to it.

Grass

The flowers of grasses always stand up high above the surrounding plants, so that they can be wind-pollinated. Each flower is very small, so you must look at it with a lens. There are three hanging anthers, and two feathery stigmas at the top of the seed box, which catch the pollen as it is blown about. There are neither sepals nor petals, but only leaf-like bracts (Fig. 97), which are called GLUMES. Several flowers are borne in a SPIKELET.

147

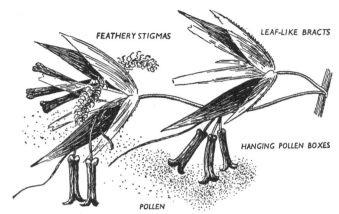

FEATHERY STIGMAS LEAF-LIKE BRACTS

HANGING POLLEN BOXES

POLLEN

Fig. 97. *Grass flowers*

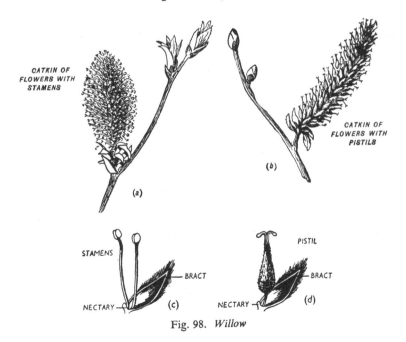

CATKIN OF FLOWERS WITH STAMENS

CATKIN OF FLOWERS WITH PISTILS

(a)

(b)

STAMENS

BRACT

NECTARY

(c)

PISTIL

BRACT

NECTARY

(d)

Fig. 98. *Willow*

148

Pollination by insects

Insects will not visit flowers unless they are attracted to them by bright colours, or by a sweet smell. The insects really come for the NECTAR, a sweet-tasting liquid, which is their food. Bees also collect the pollen (see page 103). The nectar is usually in a little bag or NECTARY which is at the bottom of the flower. The insects suck up the nectar with their tongues. While they are collecting the nectar, pollen grains which have rough surfaces stick to their hairy bodies. The pollen on the insect's body comes off on the stigma of the next flower that it visits.

The willow

Willow flowers, like the hazel, are open in the spring before the leaves appear. They do not have brightly coloured petals, but nectar with a sweet scent to attract the insects, which are chiefly bees. The stamens and the pistils are found in different flowers growing on separate trees so the plants are said to be DIOECIOUS. Plants are MONOECIOUS when the stamens and the pistils are on the same plant, whether they are present in the same or different flowers. On each tree the flowers grow together in catkins, the 'golden willow' having flowers with stamens, and the 'silver willow' having flowers with pistils. Remove a single flower from each kind of catkin and compare it with Fig. 98. The willow obviously has to be cross-pollinated as the stamens and pistils are on different trees.

The coltsfoot

The coltsfoot is a very early spring flower. It is not really one flower, but hundreds of tiny flowers called FLORETS growing massed together in a 'head' or CAPITULUM, to make a splash of colour which the insects can easily see. The capitulum is protected on the outside by a number of green bracts which together form an INVOLUCRE. Look at the capitulum and you will see that there are two kinds of flowers. The outer RAY florets are different from the inner DISC florets. Look at one of each kind of flower and compare them with Fig. 99. The outer flowers have five long, thin petals which are joined in such a way that they look like one petal. There is a pistil which has a divided stigma. The hairs on the top of the seed box are really the sepals. The inner flowers also

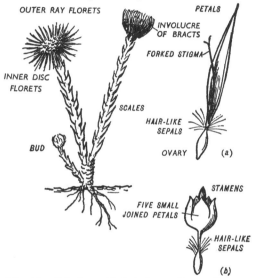

Fig. 99. *Coltsfoot.* (*a*) *Ray floret;* (*b*) *disc floret*

have hairs instead of sepals, and they have five very small petals which are joined to form a small bell. There are five stamens, and a pistil which usually does not develop. The pistils of the outer flowers are ripe before the stamens of the inner ones, so that they can only be pollinated with pollen which has been carried by insects from another plant. The capitulum closes in wet weather to protect the pollen.

The Dandelion

Like the coltsfoot, the dandelion consists of hundreds of tiny florets which all have hairs instead of sepals. Each floret has five petals which are joined to look like one single petal. The stamens are joined together by their anthers to form a hollow tube, through the middle of which the hairy pistil

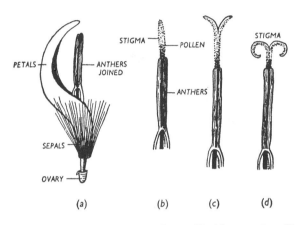

Fig. 100. *Dandelion.* (*a*) *A single flower;* (*b*)–(*d*) *stages in pollination*

grows. The ovary contains a single ovule in the coltsfoot and in the dandelion. In the coltsfoot we noticed that the pistils ripened before the stamens, but in the dandelion the opposite is true. In both cases the flowers are pollinated with pollen from another flower. In case cross-pollination does not occur, the dandelion is able to pollinate itself. The pollen boxes burst, on the inner side of the tube, as the stigma is growing up above the stamens. Pollen collects on the outer part of the young stigma and on the style (Fig. 100*b*). This can be seen if you look at the pistil under a microscope. When the stigma is above the stamens it splits in two, exposing two halves

151

which are sticky (Fig. 100*d*). Pollen grains can only grow on this sticky surface. Insects covered with pollen from another plant will touch the stigmas before they touch the stamens, and so the flower is cross-pollinated. Pollen may be carried from one floret to another in the same capitulum. If, however, it is not cross-pollinated, the two halves of the stigma curve downwards and gather pollen from its own style (Fig. 100*d*).

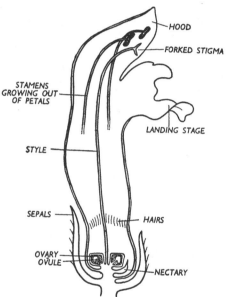

Fig. 101. *Section through a deadnettle flower*

The deadnettle

The flower of the deadnettle has sepals, petals, stamens and a pistil. The five petals are joined together to form a tube, but one petal grows out to form a 'landing stage' for the insect, and two others form a 'hood' which protects the four

152

stamens and forked stigma which are just underneath the hood. The stamens grow out of the petals. They ripen first. The ring of hairs inside the corolla tube prevents small insects from taking the nectar. When a bee alights on the landing stage of a young flower, its pushes its head into the flower to reach the nectar. In so doing it brings its back into contact with the projecting anthers. This pollen is then taken by the bee to another flower. Meanwhile the young pistil, with its forked stigma closed, is pressed against the top of the hood. Later the style projects below the anthers and the forked stigma opens to pick up the pollen from a bee's back when it next visits the flower.

The arum lily

The arum lily has a peculiar way of making sure that its flowers are cross-pollinated. It is not really one flower, but hundreds of tiny flowers which grow on the bottom half of a thickened stem. The top half of this stem is fleshy and a reddish purple in colour, and is often called a 'red-hot poker'. This central fleshy stem is protected by a large green bract which forms a tube around the flowers to protect them (Fig. 102). Remove this bract and you will see that there are three types of flowers. At the bottom of the stem there are the flowers with pistils which consist only of an ovary which contains two ovules. Above these flowers there are the flowers which contain the stamens. These flowers consist only of two or four stamens which have no filaments. Above these there are sterile flowers with neither stamens nor pistils. They have long hairs which block the tube formed around the flowers by the bract. The pistils always ripen before the stamens. Insects are attracted by the smell and the colour of the 'poker'. They are able to walk past the hairs into the

153

tube, but when once inside, they cannot get out again as the hairs block the way. The insects probably have pollen on their backs from another arum lily; if they have, they will pollinate the pistils as they crawl over them. Although trapped, the insects do not starve, as they eat the nectar which is given out by the stigmas. Later on the stamens ripen and the hairs wither. The insects can now escape, and as they walk over the stamens, pollen collects on their bodies. This pollen they may carry to another flower. Arum lilies may be found growing under hedgerows during May.

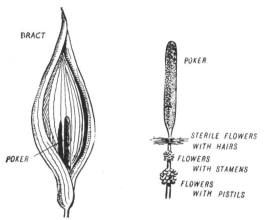

Fig. 102. *Wild arum lily*

Flower study

You will find it very interesting to make a collection of wild flowers. Press them between blotting paper until they are stiff, before putting them into a book. Start your collection in the spring and note when and where each type of flower is found. Some of the books listed in Appendix C will help you to identify your specimens.

Seeds and fruits

When the ovules are fertilized they develop into SEEDS, and the ovary enlarges to protect them. When fully ripe the enlarged ovary, with its one or more seeds enclosed, is called a FRUIT. In some fruits, such as the hawthorn and the strawberry, the receptacle becomes fleshy and forms part of the fruit. The sepals, petals and stamens die and may remain attached to the fruit or fall off. Fruits are not always juicy, sometimes they are hard, such as the nuts of hazel (Fig. 96) and acorns. Other fruits are very dry, such as lupin pods, marigold seeds, dandelion seeds, etc. During September and October look for fruits of all kinds and see how many you can find.

Seed dispersal

If all the seeds from one plant fell on to the ground below, the young plants, called SEEDLINGS, would grow so closely together that they would not be able to feed or breathe, and so they would die. The seeds must be scattered away from the parent plant. WIND, WATER and ANIMALS help to carry fruits and seeds away, other seeds are shot out by the sudden bursting of the dry ovary. These fruits are called EXPLOSIVE fruits.

Wind

Fruits that are scattered by wind are dry and very light, and have either a scale-like wing or hairs attached to them to increase the surface so that the wind can easily blow them away. In the sycamore, ash and elm, the ovary wall or PERICARP is drawn out to form the wing, whilst in lime and hornbeam the wing is formed from a bract (Fig. 103). The

155

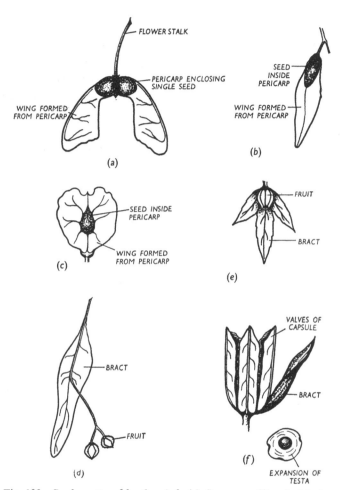

Fig. 103. *Seeds scattered by the wind.* (a) *Sycamore;* (b) *ash;* (c) *elm;*
(d) *lime;* (e) *hornbeam;* (f) *gladiolus*

fruit of gladiolus splits to set free many seeds which have wings that are formed by an expansion of the seed coat or TESTA.

After a dandelion flower has been pollinated, and the ovule has been fertilized, the petals and stamens wither, but the hair-like sepals remain attached to the tip of the ovary

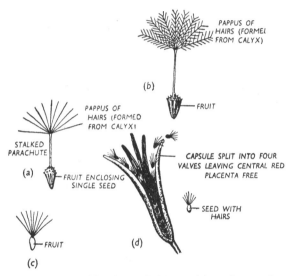

Fig. 104. *Seeds scattered by the wind.* (*a*) *Dandelion;* (*b*) *goatsbeard;* (*c*) *groundsel;* (*d*) *willow-herb*

which encloses a single seed. As the seed ripens, a stalk grows carrying the hairs away from the fruit. The PAPPUS of hairs acts as a parachute. Try to find other fruits that are like this. In groundsel and thistle the hairs remain attached to the tip of the ovary (Fig. 104).

In some plants such as willow-herb, cotton and willow, it is not the fruits which have the hairs, but the seeds. The

157

ovary wall forms a special type of fruit called a CAPSULE, which splits into several VALVES when it is ripe, to release the hairy seeds.

Many plants have capsules which contain a large number of very small seeds. When the seeds are ripe the capsule either splits into valves, or tiny holes called PORES appear. The seeds cannot fall out of the ovary unless it is blown about by the wind, so these fruits are usually on long stalks. This method of seed dispersal is called CENSER MECHANISM (Fig. 105).

Fig. 105. *Censer mechanism.* (*a*) *Snapdragon;* (*b*) *poppy*

Water

The fruits of plants living in or near to moving water have seeds that will float. These seeds may be carried a long distance by the water. Coco-nuts are specially made so that they will float in water. They have an outer, waterproof layer beneath which are many fibres which make them light enough to float.

Animals

Animals scatter seeds in many ways. Some fruits, such as goosegrass, have hooks on them which cling to an animal's coat. You will find goosegrass climbing, by means of decurved hairs, in a hedgerow. The fruits look like tiny green balls,

158

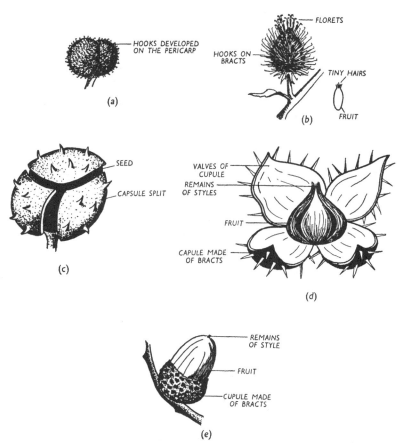

Fig. 106. *Seeds scattered by animals.* (*a*) *Goosegrass;* (*b*) *burdock;*
(*c*) *horse chestnut;* (*d*) *sweet chestnut;* (*e*) *acorn of the oak tree*

less than a quarter of an inch long. Look at the BURS on a
burdock plant. You will see that it is a capitulum similar to
that of the dandelion. Each floret produces a single-seeded
fruit with very short hairs. Each bract of the involucre surrounding the florets ends in a hook. The whole capitulum

159

may cling to the fur of an animal. It may be carried some distance before the individual fruits fall out, or before the whole capitulum drops or is rubbed off.

The seeds of horse chestnut grow inside a green, prickly case which is the ovary wall. This splits open when it falls off the tree, setting free the seeds. The prickly case of the sweet chestnut and of the beech is not formed from the ovary wall, but from a number of bracts which together form the case or CUPULE. When ripe this falls to the ground and splits open, releasing the fruits which are inside. Look closely

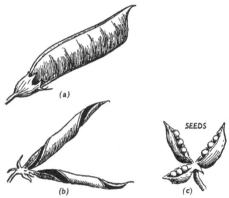

Fig. 107. *Seeds scattered by explosion.* (a) *Vetch before splitting;*
(b) *vetch after splitting;* (c) *pansy*

at these and you will see the remains of the style on the top of each one. This shows that they are fruits and not seeds (Fig. 106).

Hazel nuts (Fig. 96) and acorns (Fig. 106) may be scattered by squirrels when they are gathering them for their winter store.

Birds eat juicy berries. They may eat the juicy part and drop the seed if it is large. You may see birds carrying cherries a short distance from a tree before eating the juicy part and leaving the 'stone', inside which is the seed. If the seeds are very small, the bird will eat them, and they may pass through the bird's body unharmed. This happens when a bird eats a strawberry. The juicy part is the receptacle which has become swollen. The 'seeds' are really tiny single-seeded fruits. Have you thought of the many ways in which you, gardeners and farmers scatter seeds?

Explosive fruits

Many dry fruits split when they are ripe, with such force that the seeds are shot out for several feet. The ripe fruits or PODS of lupin, gorse, broom, pea and vetch suddenly burst open into two valves which become twisted and the seeds are scattered. The fruit of a pansy is a capsule which splits into three valves. Each valve contracts to flick out the tiny seeds.

The structure of seeds

When you first study the structure of seeds, it is wise to choose large seeds, which can be seen very easily. We shall use pea, maize and marrow seeds as they are easily obtainable and can be grown with little difficulty. Other seeds may be used and their structure compared with that of the seeds described here.

Structure of a pea seed

If you have dried peas, soak them in water for 24 hours to give them time to swell to their former size.

Now look at a pea seed. You will see a small scar or HILUM where the seed was joined to the pod. The seed is covered with a thin coat called the TESTA. Through the testa you will see the triangular-shaped young root or RADICLE. Gently press the soaked seed and a drop of moisture will come out of a tiny hole called the MICROPYLE which lies between the hilum and the tip of the radicle (Fig. 108 *a*). Remove the testa carefully and you will see the radicle very clearly. It lies between the two fleshy seed leaves or COTYLEDONS, which contain the food for the young plant. Carefully pull apart

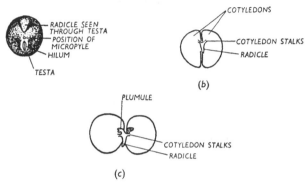

Fig. 108. *Structure of pea seed.* (*a*) *External view,* (*b*) *with testa removed;* (*c*) *cotyledons pulled apart*

the two cotyledons and you will see the young shoot or PLUMULE (Fig. 108 *c*). The tiny plant is joined to the cotyledons by two short COTYLEDON STALKS. The radicle, plumule and cotyledons are together called the EMBRYO.

Structure of a marrow seed

Marrow seeds grow inside a very fleshy fruit. Use fresh seeds if possible, or soak dry seeds.

The testa is whitish in colour and is opaque, and it must be removed to see the underlying parts. The micropyle is on one side of the seed, but it is not easily seen. Remove the testa and you will see two thin, flat cotyledons which are joined together at the base by the small, triangular radicle.

To see the plumule you must pull apart the two cotyledons (Fig. 109 b). The cotyledons store the food for the young plant. In some plants, such as the sycamore, the cotyledons are very long and thin and are folded inside the testa.

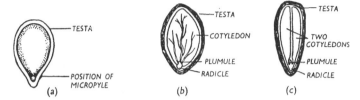

Fig. 109. *Structure of marrow seed.* (a) *External view;* (b) *internal view;* (c) *longitudinal section*

Structure of a maize fruit

The pea and marrow seeds each have two cotyledons, but the seeds of some plants, such as the maize, have only one cotyledon. The maize fruit contains only one seed whose testa is fused to the ovary wall or pericarp. The maize seed cannot be separated from the pericarp, so we must look at a maize fruit. You will see the place where the fruit was joined to the cob. The micropyle is covered by the pericarp, so you will not see it. On one side of the fruit there is a whitish patch. This is the embryo. Above it you will see a tiny scar which is the remains of the style (Fig. 110a). If the maize fruits have been soaked you may be able to see the position of the plumule and the radicle. Cut the fruit longitudinally at right angles to the white patch, look at the section with a lens and compare it with Fig. 110b. The maize fruit consists chiefly of food which is not stored in the cotyledon. Instead it forms the ENDOSPERM, which is hard and yellow on the outside but white and softer where it adjoins the embryo. The single cotyledon curves over the plumule and the radicle, separating them from the endosperm. The radicle is protected by a root sheath or COLEORHIZA, and the plumule by a sheath of leaves, the outermost of which is called the COLEOPTILE.

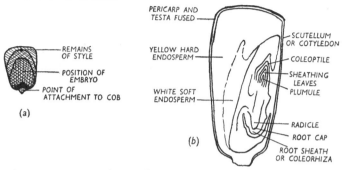

Fig. 110. *Structure of maize fruit. (a) External view; (b) longitudinal section cut at right angles to the white patch*

Germination

Seeds usually rest for a time before they begin to grow or GERMINATE. Some seeds rest for a few days, others rest until the following spring. Seeds cannot germinate without water, but many seeds remain alive although they have no water, and will germinate as soon as water is given to them. Wheat and barley will remain alive for ten or more years.

All seeds do not grow in the same way. Set as many kinds of seeds as you can and watch them growing. You cannot watch the seeds growing if you set them in soil or fibre, so set your seeds in the following way.

Large seeds. Line a glass jam-jar with blotting paper, moisten the blotting paper and place the seeds between the glass and the blotting paper. There should be half an inch of water in the bottom of the jar to keep the blotting paper moist

(Fig. 111). Very large seeds, such as the horse chestnut, may need some support, so fill up the middle of the jar with sawdust, which will hold the blotting paper against the jar.

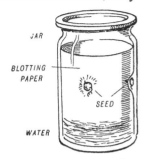

Fig. 111. *Apparatus for growing large seeds*

Small seeds. Take two small pieces of glass which are equal in size. Place a double piece of blotting paper on one piece of glass, and then arrange your tiny seeds in a row along the blotting paper (Fig. 112). Put a thin piece of wood or a wooden spill at each end so that the seeds will not be crushed, and place the second piece of glass on top. Place a rubber band or piece

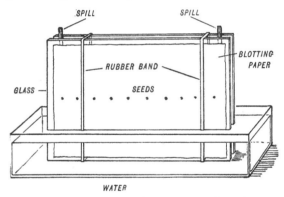

Fig. 112. *Apparatus for growing small seeds*

of string round it to hold it together and stand the whole thing in a dish of water.

Draw your seeds at different stages in their germination. You will notice that most of them germinate in a similar manner to the pea, maize or marrow seeds.

165

Germination of pea, marrow and maize

Look at Figs. 113, 114 and 115 and compare them with your diagrams. In all seeds the radicle grows first to anchor

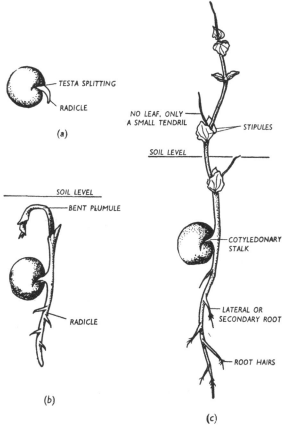

Fig. 113. *Germination of pea seed*

the plant in the soil and to take in water and minerals from the soil. The tip of each root is protected by a ROOT CAP. The radicle of the maize bursts through the testa and pericarp, but in most seeds the testa splits as the seed takes in water, freeing the radicle which grows rapidly downwards. The radicle forms the main root in the pea and the marrow, and out of it lateral or secondary roots grow. In the maize more

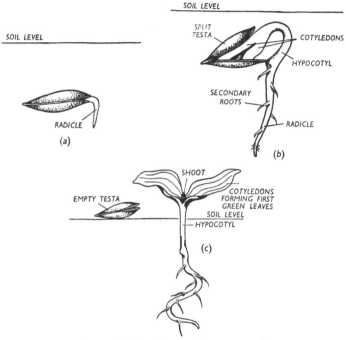

Fig. 114. *Germination of marrow seed*

roots grow from the base of the plumule, and after a short time it is difficult to tell the radicle from the other roots which are called FIBROUS ROOTS. They are also called AD-VENTITIOUS roots because they do not grow out of a root.

The cotyledons of the pea and maize seeds remain in the ground and are called HYPOGEAL cotyledons, whilst those of the marrow come above the ground and form the first green leaves. These are called EPIGEAL cotyledons.

The portion of the stem just below the cotyledon stalks,

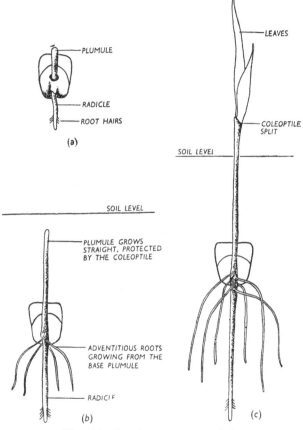

PLUMULE

RADICLE

ROOT HAIRS

(a)

LEAVES

COLEOPTILE
SPLIT

SOIL LEVEL

SOIL LEVEL

PLUMULE GROWS
STRAIGHT, PROTECTED
BY THE COLEOPTILE

ADVENTITIOUS ROOTS
GROWING FROM THE
BASE PLUMULE

RADICLE

(b)

(c)

Fig. 115. *Germination of maize fruit*

which is called the HYPOCOTYL, elongates rapidly upwards in the marrow, taking with it the seed. The tip of the shoot is protected by the cotyledons and by the testa as it pushes its way through the soil. The empty testa case falls off when the shoot is above the ground, and the cotyledons form the first green leaves. As the pea and maize cotyledons remain in the ground the tip cannot be protected by them or by the testa. The cotyledon stalks of the pea increase in length carrying the plumule out of the seed. The plumule remains bent to protect the tip until it has passed through the soil. The maize plumule grows out through the skin and is straight as it pushes its way through the soil. The tip is protected by sheathing leaves. The outermost sheath, or COLEOPTILE, has a hard, pointed tip which is not damaged as it pushes its way through the soil. The inner leaves burst through the coleoptile when it comes above ground.

Conditions necessary for germination

Set up five jars with wet blotting paper, as described on p. 164, but put the same kind of seeds in each jar. Pea seeds could be used. Place one jar on the window-sill in the light. The seeds in this jar will have light, warmth, air and water. Place the second jar in the dark, and you will then be able to tell whether seeds germinate better in the light or in the dark. Place the third jar outside where it is cold, and the fourth jar close to a radiator. This will show you whether temperature affects the rate of germination. In the fifth jar put seeds which have been killed by boiling them or by soaking them in formalin. Set up a sixth jar but do not wet the blotting paper; then the seeds will not have any water. A seventh jar must be filled to the top with cold, distilled water and covered over so that the seeds will not have any oxygen. Set up an eighth jar in the normal manner, but place the seeds so that the radicles are pointing in different directions.

Carefully note the results of your experiment and you will

169

find that seeds grow best when they have WATER, OXYGEN and NORMAL TEMPERATURE. Seeds must be ALIVE, as the dead seeds will not grow. You will notice that the radicles grow downwards regardless of the direction in which they are pointing when planted. Many seeds germinate equally well in the light or in the dark although they are usually in the dark in the soil. As the shoots grow bigger, however, you will notice that the shoots in the dark become straggly and have no green colouring matter or CHLOROPHYLL. Tomato seeds will not germinate in the light.

Sow some very large seeds, with and without their cotyledons, and some very small seeds (*a*) very deeply, (*b*) at the surface of the soil. All the seeds will germinate, but the small seeds and the large seeds without cotyledons which were sown deeply die before they reach the surface of the soil. This is because a seed must have enough food stored inside it to enable the shoot to grow above the soil. Plants are not able to make their own food until the green shoots are above the soil (see Book II). The food stored in seeds must be changed into a soluble form, by substances called ENZYMES which are present in the seeds, before it can pass to the radicle and the plumule. Soluble food from the cotyledons passes through the cotyledon stalks to the radicle and plumule in the pea and marrow. The cotyledon in maize gives out enzymes which make the food in the endosperm soluble. This is absorbed by the cotyledon which then passes it to the radicle and the plumule.

CHAPTER 7

THE STRUCTURE OF A FLOWERING PLANT

Although nearly all the plants with which you are familiar have ROOTS, STEMS, LEAVES, FLOWERS and FRUITS, there is a great variety in their structure.

Roots

When you were growing your seeds you noticed that the roots always grow first. This is very necessary because without roots to hold the plant firmly in the soil, the plant would be easily blown over. Roots also have to take in the water and mineral salts that the plant needs.

Look at the root of a germinating pea or bean seed (Fig. 116a). You will see that it has a MAIN ROOT formed from the radicle, out of which grow the SIDE or LATERAL ROOTS. This is called a TAP ROOT SYSTEM. Look now at the roots of a germinating maize seed and you will not be able to distinguish the radicle from the later formed roots which grow from the base of the plumule. This is called a FIBROUS ROOT SYSTEM (Fig. 116b).

The tip of all roots is protected by a ROOT CAP while it is pushing its way through the soil. The growing region of the root is immediately behind the tip. This can be shown by marking lines, with cotton dipped in Indian ink, one-tenth of an inch apart from the tip along the root of a seedling.

As the root grows, the marks immediately behind the tip get farther apart than those nearer to the seed and there are no marks on the newly formed part of the root (Fig. 117). Look again at a young root and you will see thread-like hairs growing out of the root just behind the tip. These hairs are called ROOT HAIRS, and it is through them that the roots

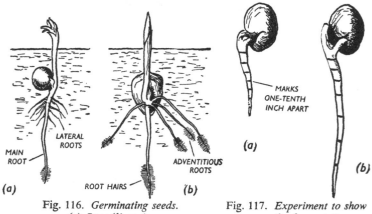

Fig. 116. *Germinating seeds.*
(*a*) *Pea;* (*b*) *maize*

Fig. 117. *Experiment to show growth of a root tip*

absorb water and dissolved minerals. As the roots grow older, the root hairs wither away and are replaced by more root hairs which grow on the newly formed root behind the tip.

Roots with special functions

You are very familiar with the swollen roots of carrots, beetroot, radishes, etc. (Fig. 118*a*). These plants live for two years (and are called BIENNIALS). During the first year roots and leaves grow, and a large quantity of food is stored in the TAP ROOT which becomes very fat. This food is used up during the second year, when more leaves, the flowers and the seeds grow. Then the whole plant dies leaving seeds to grow into new plants.

In a few plants such as the dahlia and the celandine, some of the fibrous roots become swollen with food and are called

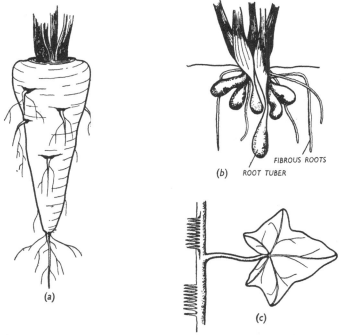

Fig. 118. *Roots with special functions:* (a) *tap root of carrot swollen with food;* (b) *fibrous roots of celandine swollen with food;* (c) *adventitious roots growing from stem of ivy*

ROOT TUBERS (Fig. 118b). The remaining fibrous roots absorb water and minerals and anchor the plant in the soil. If you look at these plants during the winter you will see that the aerial parts have died, leaving only the tubers in the ground, from which the new plants grow the following year.

ADVENTITIOUS ROOTS are those which grow out of stems and leaves. Place twigs of willow or poplar in a jar of water and leave them for a time. You will see adventitious roots growing out of the stem. Look at the stem of IVY (Fig. 118 c) and you will see small adventitious roots growing out of the stem. These roots stick to the bark of a tree or to the wall up which the ivy is growing.

Stems

Some low-growing plants, such as the daisy, do not have stems. The leaves are arranged like a rosette so that they are all in the light. These plants always grow in the open. In the majority of plants leaves grow out of the stem. The angle which the leaf makes with the stem is called the AXIL. A bud is usually borne in the axil and is called the AXILLARY BUD. That part of the stem from which a leaf grows is called a NODE, and the part of a stem between two nodes is called an INTER-NODE. The stem holds the leaves and the flowers up to the light. Stems are usually strengthened so that they can hold up all the shoots. If the stems are too weak to hold up the shoots, the plants climb up other plants. These CLIMBING PLANTS, as they are called, climb in different ways (Fig. 119). The convolvulus, hop and the bean TWINE their stems around any plant or stick that is near to them (Fig. 119 a). Find as many twining plants as you can, and you will see that some twine in a clockwise manner like the hop, and others twine in the opposite direction like the convolvulus. The ivy climbs by means of its ADVENTITIOUS ROOTS (Fig. 118 c). The rose and the blackberry have HOOKS which grow out of the stems (Fig. 119 b). These hooks enable the plants to SCRAMBLE over other plants. Some plants climb by means of TENDRILS, which may be modified stems (as in the white bryony, Fig. 119 c),

174

modified leaves as in the yellow pea, or modified leaflets as in the garden pea (Fig. 138). Make a collection of climbing plants and notice how they climb.

The water, containing dissolved mineral salts, which is taken in by the root hairs, passes through special conducting channels in the root and stem. Place a piece of plant stem in a jar of red ink and leave it for some time. The red ink

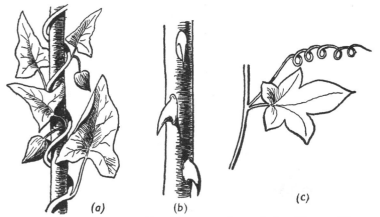

Fig. 119. *Climbing stems: (a) twining stem of convolvulus; (b) scrambling stem of rose, with hooks; (c) white bryony with stem tendrils*

rises up the conducting channels, which become stained and so are clearly seen. If you stand a white dahlia flower in red ink, you will see the colour passing along the veins in the petals.

Stems with special functions

Food is often stored in the stems of plants. A potato is part of an underground stem very swollen with food which is chiefly starch. Potatoes are STEM TUBERS (Fig. 120). Look at a potato and you will see the 'eyes'. These are really BUDS

175

which are growing in the axils of tiny scale leaves. These buds tell us that the potato is a stem, because roots do not have buds. The potato is covered with a layer of cork which is pierced by tiny holes called LENTICELS through which the potato breathes. You can see these small dots quite easily on a fairly new potato. When you set a potato in the ground

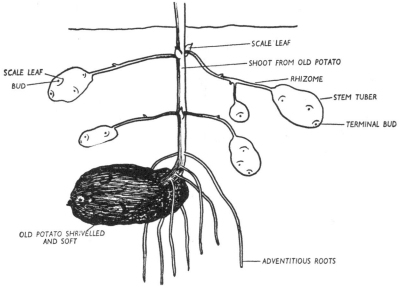

Fig. 120. *A potato plant*

a shoot grows first from one of the 'eyes', and from the base of the shoot ADVENTITIOUS ROOTS grow. Later, more shoots and roots may grow from the other eyes. The food in the old potato is used to feed the growing plant, and so the potato shrinks and becomes soft. If you look at the lower part of the stem you will see tiny scale leaves, from the axils of which stems grow farther down into the ground (Fig. 120). These stems are called rhizomes or stolons. Food which is made in

176

the leaves (see Book II) is passed down to and stored in the ends of these stems which gradually swell to form the STEM TUBERS. When the plant dies down in the autumn, the new

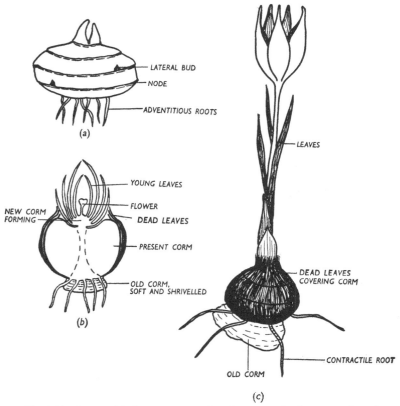

LATERAL BUD
NODE
ADVENTITIOUS ROOTS
(a)

YOUNG LEAVES
NEW CORM FORMING
FLOWER
DEAD LEAVES
PRESENT CORM
OLD CORM, SOFT AND SHRIVELLED
(b)

LEAVES
DEAD LEAVES COVERING CORM
CONTRACTILE ROOT
OLD CORM
(c)

Fig. 121. *Corms.* (*a*) *Crocus corm with scales removed;* (*b*) *section through crocus corm to show position of new corm;* (*c*) *crocus plant*

potatoes become detached and can each grow into a new plant the following year.

Crocus CORMS are also stems that are swollen with food (Fig. 121). The base of the flowering stem swells up as it

177

receives food from the leaves (Fig. 121 *b*). When the leaves die their bases remain attached to the corm protecting it. If you remove these, you will see that each one grows out

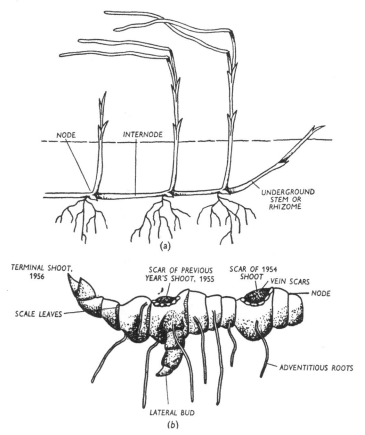

Fig. 122. *Rhizomes.* (*a*) *Couch grass;* (*b*) *Solomon's seal*

of a node (Fig. 122 *a*). Tiny lateral buds grow in the axils of these leaves. Cut through the larger, terminal bud and you will see the young leaves and flower. As these grow they use up the food which is stored in the corm. This corm gets smaller. The leaves make food which is passed down to and stored in the stem at the base of the leaves and flower. This grows larger forming a new corm on top of the old one. The new corm is pulled down in the ground by some of its roots which are contractile. The corms of arum lily and montbretia persist for several years, so you will see several corms growing on top of one another.

Some plants have stems, called RHIZOMES, which grow horizontally in the soil. They have nodes and internodes and may or may not be swollen with food. Couch grass has a rhizome which is not swollen with food (Fig. 122a). Adventitious roots and shoots grow from the nodes. If a piece of rhizome is left in the ground, shoots and roots will grow from it, and so couch grass is a weed that is very difficulty to destroy. Try to find other plants that have similar rhizomes.

The rhizomes of iris and Solomon's seal are swollen with food. You can distinguish the nodes and the internodes on each rhizome and both have adventitious roots growing out of them. The leaves of the iris grow from the nodes in two parallel rows. The bases of the leaves completely cover the rhizome. As the leaves die their remains can be seen attached to the rhizome, and when they finally rot away you will see marks where the veins entered the leaves. The rhizome of Solomon's seal sends up a terminal shoot each year which bears leaves and flowers. A single round scar is left on the rhizome when the shoot dies. Vein marks can be seen in the scar. By counting the number of scars you can tell the age of the rhizome. It is covered with small scale-leaves (Fig. 122b). The iris rhizome ceases to grow when a flower is formed and growth is continued by two lateral buds (Fig. 123).

179

In Solomon's seal a lateral bud grows out from below the terminal shoot to continue the growth of the rhizome. As this grows downwards at first, the rhizome always grows in

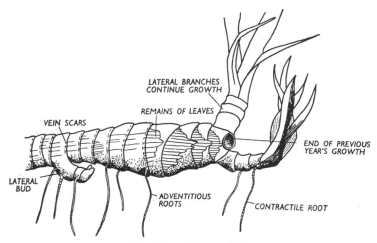

Fig. 123. *Rhizome of iris*

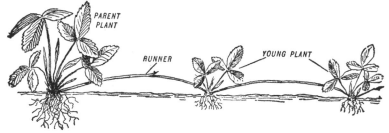

Fig. 124. *Strawberry runner*

the soil. The iris rhizome is pulled down into the soil by contractile roots. Lateral buds may grow in both rhizomes to spread the plant. As the rhizome gets old, it rots, and these branches may be separated from one another. The main stem of some plants such as creeping Jenny creeps along the ground giving off adventitious roots from the nodes. These stems are called CREEPING STEMS. Strawberry plants send out lateral branches or RUNNERS above the ground. These have

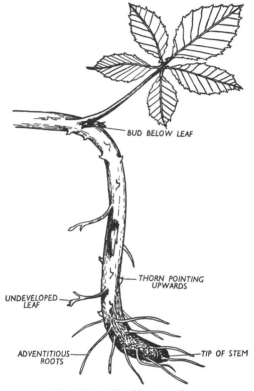

Fig. 125. *A blackberry stolon*

very long internodes and small scale-like leaves growing from the nodes. The tips of these runners produce new plants which in turn produce more runners (Fig. 124). The runners eventually die and the new plants grow quickly. If the branches of blackberry and gooseberry bushes touch the ground the

181

tip swells and numerous adventitious roots grow out of it. Leafy shoots grow out from the axils of the small leaves which are intermingled with the roots above the ground. These stems are called STOLONS (Fig. 125). It is easy to recognize a blackberry stolon because the thorns point upwards instead of downwards and the buds are below the leaves and not above them. THORNS are modified branches which protect the plant from animals. The hawthorn tree has thorns.

Duration of plants

Many plants, like peas and beans, live for one season and then die, leaving seeds to grow into new plants the following year. These plants are called ANNUALS. We have learned that BIENNIALS live for two years. The majority of plants in this country, however, live for several years and are called PERENNIALS. In HERBACEOUS perennials like Chrysanthemums, the shoots above the soil die in the autumn leaving only roots, stems and buds in the ground. The buds shoot up the following spring. The aerial parts of TREES and SHRUBS continue to grow every year, and do not die. These are called WOODY PERENNIALS. DECIDUOUS plants lose their leaves in the autumn. EVERGREEN plants do not lose their leaves.

Growth in thickness of stems

As the aerial parts of trees and shrubs continue to grow every year, the stems have to grow wider to bear the weight of the growing branches. The greenish-coloured layer just under the BARK consists of cells which grow and divide to make the stem or trunk wider. New wood is formed which carries water and dissolved minerals from the roots to the leaves. This new wood is called SAP WOOD. As the sap wood gets older it gradually dies, forming a central dry mass of dead cells which is the real support of the tree. This central wood is called HEART WOOD and is a different colour from the sap wood. The cells of wood formed in the spring are larger than those formed in the autumn. So a clear ring of wood is added each year. These ANNUAL RINGS can easily be seen in a section of a tree. KNOTS in wood are sections across branches which have gradually become enclosed by the growth in thickness of the stem. Between the new wood and the bark new cells are formed which carry food from the leaves to the other parts of the plant. If you cut off the bark of a tree you usually cut off these cells as well, so killing the tree. This important part of the trunk is protected by the BARK which is made up of cork through which water cannot pass. As the trunk gets wider the bark, which cannot grow, either splits (elm tree) or comes off in flakes (plane tree). New bark is then formed. Old roots are also covered with bark. Look at the trunks of different trees and try to recognize the trees by their bark (see Appendix B).

Buds

Look at the twigs of any tree that loses its leaves at the beginning of winter, and you will see a number of BUDS (Fig. 128). A bud is a short shoot with a number of young LEAVES on it closely packed together. At the end of every twig there is a bud, called the TERMINAL BUD. In addition, there is a bud at the base of every leaf on the plant, between the leaf stalk and the stem of the twig. These buds are called AXILLARY BUDS because they grow in the AXILS of the leaves (Fig. 128).

The young leaves inside the buds are extremely delicate. When they are first forming they require protection from too

183

much heat, cold, dryness, and moisture, as well as against the
attacks of animals. Buds are protected in various ways. They
are often protected by the stalk of the leaf until it falls off the

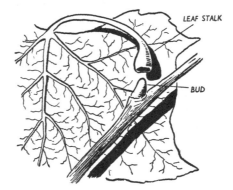

Fig. 126. *Bud of plane tree protected by leaf stalk*

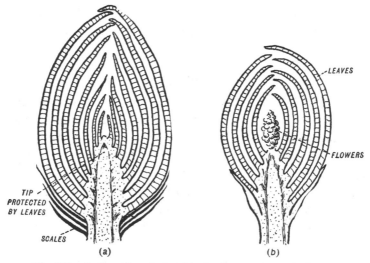

Fig. 127. *Buds cut lengthwise:* (a) *showing growing end of stem;*
(b) *showing young flowers*

184

tree. In the plane each bud is completely covered by the hollow end of the stalk (Fig. 126).

Horse chestnut buds are large, dark-coloured and sticky (Fig. 128). This gum which makes the buds sticky is only on the outer surface of the bud and protects the young leaves from insects and other animals. Carefully pull a horse chestnut bud to pieces. On the outside are small leaves called SCALE LEAVES, which are packed closely together and are waterproof. The spaces between the young leaves themselves are filled with hairs which protect the young leaves from the cold, and also prevent them from losing too much moisture and so drying up.

Cut a bud in two lengthwise and you will see a very short stem with many small leaves growing out of it (Fig. 127 a). In the spring the end of the stem, in the middle of the bud, will lengthen, and the green leaves will expand and grow. No cap protects the growing end of the stem as it did in the root, because the stem does not have to push its way through

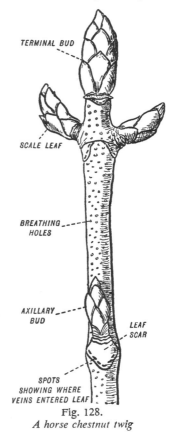

Fig. 128.
A horse chestnut twig

anything hard such as the soil. The delicate growing end of the stem is well enough protected by the young leaves which are neatly folded over it (Fig. 127 a).

185

Some of the terminal buds, when cut open, show the beginnings of a flower as well as leaves (Fig. 127*b*). The warmth of spring causes the buds to open. Since the leaves and flowers are already formed inside the bud, they soon grow out.

Twigs

Twigs are small branches of trees. Look at a horse chestnut twig during the winter (Fig. 128). Beneath each bud there is a horse-shoe shaped scar, which shows where a previous year's leaf was joined to the stem. These scars are called LEAF SCARS. On the scar there are a number of small spots, usually five or seven in the horse chestnut, which mark the places where veins from the stem entered the leaf.

If a terminal bud grows into a short stem with flowers at the end of it, it then stops growing. Later, when the flowers and fruits have fallen, the short stem which bore them drops off, leaving a scar which is called the INFLORESCENCE SCAR. Growth of the twig is continued by the next pair of axillary buds, so making a fork in the twig, in the middle of which you will see the scar of the flower stem (Fig. 129).

On the thin bark you will see many small spots. These are BREATHING HOLES or LENTICELS. The bark is a dead covering, but the living cells underneath must breathe. This they do through these holes. During the winter the lenticels are closed.

Keep some twigs in water in a warm room and watch the buds opening. You will notice that the scale leaves fall off as the bud opens, leaving a ring of scars on the stem. The ring of scars made by the scale leaves of the previous year's terminal bud can be seen lower down the stem. In one year the stem has grown from this ring of scars to the terminal bud. As these

rings denote a year's growth we can call them GROWTH RINGS.
By counting the number of rings the age of a twig is found.
Twigs vary very much in the amount they grow in one year.
Some twigs grow several inches, or even several feet, as for
example lime trees. Others may grow only a fraction of an

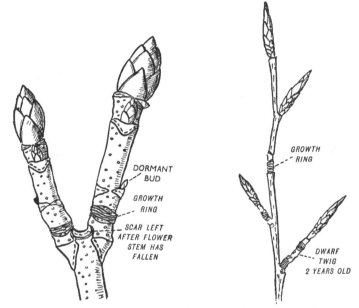

Fig. 129. *Horse chestnut showing fork made
by growth of axillary buds after the flower
stem has fallen*

Fig. 130. *Beech twig*

inch in one year. Look at the small side branches on a
beech twig (Fig. 130). Although very short they are probably
several years old. These small branches are called DWARF
TWIGS.

Side branches usually grow more slowly than the main
branch, but if the main branch is cut, the side branches grow

more quickly. If you have a hedge of privet, or of any other plant, you must repeatedly cut off the terminal branches so that the side branches will grow more quickly and produce a thicker hedge. Some lateral buds remain very small and are called DORMANT BUDS. They grow only if the stem is broken above them.

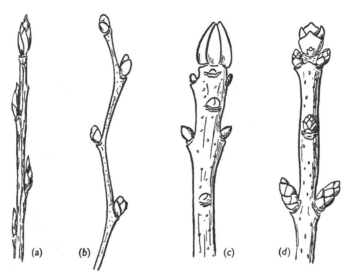

Fig. 131. *Twigs.* (a) *Poplar;* (b) *lime;* (c) *ash;* (d) *sycamore*

The recognition of common twigs

Trees can be recognized during the winter by their twigs. The bark of twigs varies in colour and the buds differ in size, shape, colour and their arrangement on the stem. Twigs of six of the most common trees are briefly described here (Figs. 129–131).

Poplar. Pointed buds, arranged alternately, and lying close to the light-brown stem.

Lime. Reddish-green stems and buds. Round buds arranged alternately down 'zig-zag' stem.

Sycamore. Greyish stems. Green buds in pairs down the stem. One terminal bud.

Ash. Grey stems which are broad and flat at the nodes. Black buds.

Beech. Light-coloured bark. Long, thin, pointed, brown buds arranged alternately. Dwarf twigs (Fig. 130).

Horse chestnut. Large, sticky, brown buds arranged in pairs down the stem. Horse-shoe shaped scars (Fig. 128).

Tree study

It is very interesting to study trees at different seasons of the year. Begin your study in the autumn at the beginning of the school year. Collect leaves and make leaf tracings (see Appendix B). Collect and draw the fruits of the trees. The elm tree fruits are usually found before the autumn. During the winter look at the twigs and draw them. Look carefully at the bark and either make bark scribbles or plaster casts of the bark (see Appendix B). Notice when the trees are in flower. Press flowers as described on page 154.

Leaves

It is necessary for all leaves to get as much air and light as possible. Leaves are arranged on the stem so that one leaf does not shade another. Sometimes they are in pairs opposite one another, and each pair is usually at right angles to the pair above, as in the privet (Fig. 132*a*). In most plants the leaves come off singly, as for example in the hazel (Fig. 133). In spite of their arrangement on the stem, the leaves on the branches at the side of the tree cannot get enough light. So

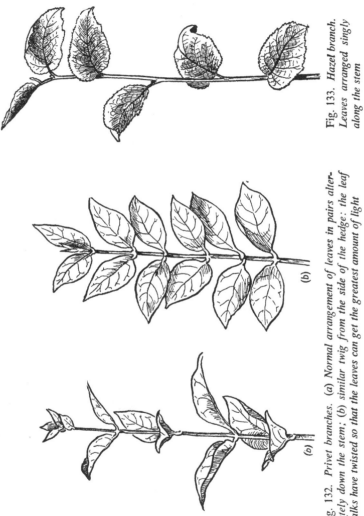

Fig. 133. *Hazel branch. Leaves arranged singly along the stem*

Fig. 132. *Privet branches. (a) Normal arrangement of leaves in pairs alternately down the stem; (b) similar twig from the side of the hedge: the leaf stalks have twisted so that the leaves can get the greatest amount of light*

(a)

(b)

the leaf stalks often twist to bring the leaves into such a position that they receive the greatest amount of light (Fig. 132*b*). The leaves of a dandelion have no long stalks to hold them out to the light. Instead its leaves spread out to form a rosette so that each leaf gets a share of the light.

The structure of a leaf

A leaf consists of two parts: (1) The LEAF STALK or PETIOLE, (2) the BLADE or LAMINA.

The leaf stalk varies in length. Leaves growing in shady places have long stalks to hold them to the light, whereas others like the dandelion have no stalks. The base of the leaf stalk is sometimes flattened and may even surround the stem to which it is attached. This is often seen in large leaves. Some leaves have leafy outgrowths from the petiole which are called STIPULES (Fig. 138).

The leaf blade varies very much in structure. A number of lines can be seen on the blade. These lines are called VEINS and are the canals along which water and food pass to the leaf, and food made in the leaf passes back into the stem. The veins also give the blade support and make it rigid.

Compare the leaf of a grass or iris (Fig. 134*d*) with that of an elm (Fig. 134*b*). You will see that the veins are very differently arranged in each. In the birch leaf there is one main vein called the MIDRIB from which small veins branch forming a network over the blade. These leaves are said to be NET VEINED. The veins in an iris leaf run parallel to one another and do not form a network. This leaf is PARALLEL VEINED.

The shape of a blade varies very much. The edge or margin of the leaf also varies. The margin of a privet leaf is smooth (Fig. 132), but an elm leaf has toothed edges (Fig. 134*b*), and an oak leaf a wavy outline.

The blade is sometimes cut up (Fig. 134c), and is not whole as it is in the privet. The leaves of the water crowfoot are interesting to look at (Fig. 1). The leaves above water are

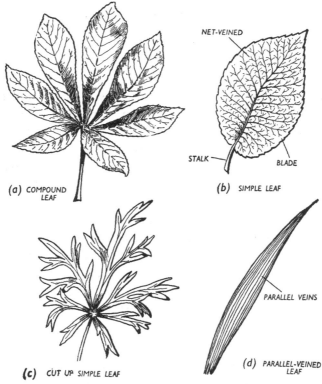

Fig. 134. *Leaves: (a) horse chestnut; (b) elm; (c) monkshood; (d) iris*

very little cut up, but the leaves in the water are cut up so much that they look like bunches of green threads. This prevents them from getting damaged in the water, since currents flow between the green threads.

As long as the divisions do not reach the midrib, the leaf is SIMPLE. When the blade is divided to the midrib a number of separate LEAFLETS are formed, and the leaf is then COMPOUND (Fig. 134a). At the base of a leaf there is always a bud. You will easily recognize a compound leaf because there will not be a bud at the base of each leaflet, only at the base of the leaf.

Strip a piece of skin from an iris leaf and look at it under a microscope. You will see a number of tiny holes, PORES or STOMATA (Fig. 135). All leaves, except those growing in water, have stomata which are usually more numerous on the lower side of the leaf. Each stoma is surrounded by two GUARD CELLS which open and close the stoma. Look at a section of a leaf under a microscope and compare it with Fig. 136. Nearly all leaves are covered with a colourless skin or EPIDERMIS which may be protected by a waterproof layer, the CUTICLE.

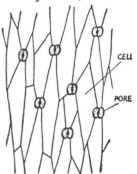

Fig. 135. *Skin of iris leaf seen under microscope*

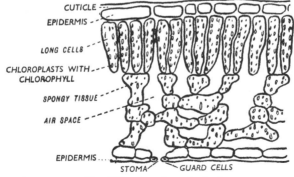

Fig. 136. *A section of a leaf*

193

Most of the green-colouring matter, called CHLOROPHYLL, is contained in grains or CHLOROPLASTS in the long cells which are beneath the upper skin. The underside of a leaf is lighter in colour than the upper surface, because the lower cells contain little chlorophyll. The guard cells are the only cells in the epidermis which contain chlorophyll. The cells in the lower half of the leaf are packed loosely together and have air spaces between them. These air spaces form passages between the cells of the upper surface and the stomata and allow gases to pass throughout the leaf. Leaves are very important parts of a plant because they breathe, give out water and manufacture food (see Book II).

Leaves with special functions

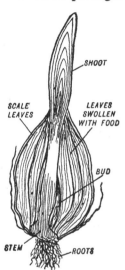

Fig. 137. *Bulb cut in two*

Leaves are sometimes modified to perform special functions. Look at an onion BULB that has been cut in two longitudinally (Fig. 137). You will see that it consists chiefly of leaves that are swollen with food which is sugar. The stem is at the bottom of the bulb, and out of it grow the roots and the fleshy leaves. In the centre of the onion you will see the terminal bud which develops into leaves and flowers. Buds grow in the axils of the fleshy leaves. These develop into new bulbs. On p. 174 we read that a few plants have to climb in order to reach the light. Some of these plants climb by means of TENDRILS which may be modified leaves (yellow vetchling) or modified leaflets (garden pea, Fig. 138). Leaves like the holly have SPINES on them to protect them

194

from animals. In the gorse whole leaves or even branches are modified into SPINES. SCALE LEAVES (see p. 185) are also modified leaves.

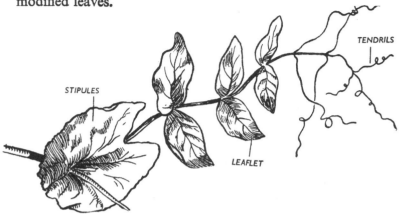

Fig. 138. *Garden pea. Compound leaf with tendrils*

Vegetative propagation

If we want new plants we usually set seeds. We can, however, plant tubers, rhizomes, corms or bulbs which will grow into new plants. From the nodes of runners and stolons new plants grow which can be detached from the parent plant. New plants may even grow from the leaves of a begonia plant. This is called VEGETATIVE PROPAGATION.

195

QUESTIONS

Chapter 1

1. Why is it necessary to put water plants into an aquarium?
2. Why are the leaves of many water plants finely divided?
3. Of what use are the winter buds?

Chapter 2

1. Explain how you would try to find (*a*) amoeba, (*b*) hydra.
2. How does (*a*) the amoeba, (*b*) the hydra move about?
3. Describe how the amoeba catches and eats its food.
4. 'The amoeba never dies.' Is this true?
5. How does the hydra catch its food?
6. Old hydras die. How are new hydras formed?
7. Why are leeches called 'blood-suckers'?
8. Why do the snails and swan mussels have shells? How are these shells formed? What is the difference between a snail shell and a swan mussel shell?
9. How does a snail move?
10. What does a snail eat and how does it eat its food?
11. How does a pond snail breathe?
12. Of what use are the siphons of the swan mussel?
13. The pond snail has eyes and feelers. Why are they absent in the swan mussel?
14. Describe the life history of a young swan mussel.
15. Why is it necessary to have water fleas and cyclops in your aquarium?
16. How do the following animals protect their young ones: (*a*) water flea, (*b*) cyclops, (*c*) fresh-water shrimp?

196

17. How do water fleas and cyclops survive during the winter or during a drought?

18. How could you recognize a fresh-water shrimp from a water slater?

19. What are scavengers and why are they so useful in an aquarium?

20. What is meant by moulting?

21. Describe the life history of one of the following animals: (a) carnivorous water beetle, (b) gnat.

22. Give reasons for the following:

(a) Covering over tank containing leeches or water beetles.

(b) Putting carnivorous water beetles and notonecta in separate tanks away from all other animals.

23. Name five pond insects which have to come to the surface of the water to breathe. How is the oxygen carried to all parts of an insect's body?

24. What does a water scorpion eat and how does it catch its prey?

25. How does a young water scorpion differ from its parents?

26. What are the chief differences between the two types of water boatmen?

27. Describe how the caddis-fly larva protects its soft body.

28. Why is it necessary to put partially submerged plants into an aquarium containing young caddis-flies and young dragon-flies?

29. Caddis-fly larvae and dragon-fly nymphs have special organs which enable them to breathe in the water. Describe these organs.

30. Describe how a dragon-fly nymph catches its food.

31. Why are gnats harmful?

Chapter 3

1. Why is a fish's body streamlined?

2. Why does a fish not have eyelids?

3. How does a fish swim?

4. How does a fish breathe?

5. Describe the life history of an eel.

6. Compare and contrast a frog and a toad.

7. Why is a frog's skin wet?

8. How does a frog catch its food?

9. Briefly describe the life history of a frog.

10. How does a young newt differ from a tadpole?

11. How would you recognize the eggs of (a) frogs, (b) toads, (c) newts?

12. What happens to frogs, toads and newts during the winter.

Chapter 4

1. Why is the skin of an earthworm damp?

2. Carefully explain how an earthworm lays its eggs.

3. Briefly describe why earthworms are useful animals in the soil.

4. What happens to earthworms when the soil is very dry?

5. What do slugs and snails eat, and how do they eat their food?

6. How can slugs and snails tell where they are going?

7. How could you distinguish a centipede from a wireworm?

8. Are millipedes and centipedes useful or harmful? Give reasons.

9. How is oxygen carried round the body in (a) earthworms, (b) centipedes?

10. How does a wood-louse breathe?

11. Wood-lice are good mothers. How do they take care of their young ones?

12. Explain what is meant by 'moulting'.

13. Briefly describe how different kinds of spiders catch and eat their food.

14. What are 'book lungs'?

15. How could you recognize (*a*) an insect, (*b*) a spider?

16. Imagine that you are watching a garden spider weaving its web. Describe what you would see.

17. At what time of the year would you look for spiders' eggs, and where would you look for them?

18. How can you distinguish a moth from a butterfly?

19. Briefly describe the life history of a butterfly or of a moth.

20. Why are caterpillars harmful? Make a list of the names of all the caterpillars that you know are harmful, and say on which plants they may be found.

21. Explain how ichneumon flies kill caterpillars.

22. What happens to the moths and butterflies during the winter?

23. Name the animals in the garden that help us to get rid of aphides and say how they do it.

24. Why are ants fond of aphides?

25. What is cuckoo-spit? Briefly describe how it is made.

26. What do frog-hoppers eat?

27. Briefly describe the life history of an aphis.

28. Are the following insects useful or harmful? Give reasons for your answer: (*a*) ladybirds, (*b*) crane flies, (*c*) hover-flies, (*d*) click beetles.

29. What is (*a*) a wireworm, (*b*) a leatherjacket?

30. Describe the life history of either a ladybird or a hover-fly.

31. What is meant by 'social insects'?

32. Describe the life in either an ant hill or a hive.

33. What are the main differences between the life in an ant hill and the life in a hive?

Chapter 5

1. Briefly describe the chief characteristics of all reptiles.

2. How could you tell a grass snake from an adder?

3. How is the body of a snake adapted to enable it to swallow its prey whole?

4. If you watch a snake you will see that it does not blink. What is the reason for this?

5. What is meant by the words oviparous and viviparous?

6. How would you recognize a lizard from a newt?

7. Compare and contrast a snake and a slow-worm.

8. Why must you not pick up lizards or slow-worms by their tails?

9. Why are birds' bones very light?

10. Briefly describe the structure of a bird's wing.

11. Of what use is the keel on the breastbone of a bird.

12. With the help of diagrams, briefly describe the structure of the feathers that are found on the wings and on the body of a bird.

13. How does a bird get a good supply of oxygen?

14. Why does a bird have bony plates in its eyes?

15. Of what use is a bird's beak? Why are the beaks of birds of different shapes? Describe three different kinds of beaks.

16. How can you tell from a bird's feet where it lives?

17. Write an essay on 'care of the young in birds'.

18. By means of a carefully labelled diagram describe the structure of an egg.

19. What are the chief characteristics of all mammals?

20. What is meant by the terms 'warm-blooded' and 'cold-blooded' animals?

21. Write a brief essay on the migration of birds.

22. What is a hoof? With the help of diagrams draw two types of hoofs and say why the animal has this particular type of hoof.

23. Compare the teeth of a hoofed mammal with those of a gnawing mammal. Say briefly why they are similar.

24. Explain what is meant by 'chewing the cud'.

25. Write a brief essay on how hoofed mammals protect themselves.

26. In what ways do the horns of cows differ from the antlers of the deer?

27. How do the teeth of a dog differ from those of a horse? Give reasons for these differences.

28. Write an essay comparing and contrasting cats and dogs.

29. Briefly describe any two beasts of prey that are wild in this country.

30. Why are rats and mice harmful animals?

31. Write a little about a baby horse and a baby mouse and give reasons why they are very different.

32. Write an essay on rabbits.

33. In what ways do hares differ from rabbits?

34. Briefly describe a squirrel, saying all that you can about it.

35. How do the following animals protect themselves from their enemies: (a) hedgehog, (b) squirrel, (c) rabbit?

36. What is meant by hibernation? Mention a few animals that hibernate and say why they do it.

37. How can you tell that a mole is present in a field?

38. Write an essay on 'how mammals take care of their young ones'.

39. How would you tell a shrew from a mouse?

40. Describe the structure of a bat's wing.

41. Are bats blind? How do they know where they are flying?

42. Describe one thing that is peculiar about bats.

Chapter 6

1. Why is the flower such an important part of a plant?

2. Describe the four parts of a flower and say of what use each part is.

3. Why must flowers be pollinated?

4. Why is cross-pollination better than self-pollination?

5. What are the chief characteristics of (*a*) wind-pollinated, (*b*) insect-pollinated flowers?

6. Describe the method of pollination of one wind-pollinated flower, and one insect-pollinated flower.

7. Describe one way in which a flower prevents self-pollination.

8. What is the difference between a seed and a fruit?

9. Explain how the structure of some fruits and seeds enables them to be scattered by wind.

10. Describe the various ways in which animals help to scatter seeds.

11. What is meant by censer mechanism and explosive mechanism in the dispersal of seeds? Describe two fruits whose seeds are scattered in these ways.

12. Compare and contrast the structure of a pea and of a maize seed.

13. Which part of a plant always grows first? Why?

14. How are the shoots of the following plants protected as they grow through the soil: (*a*) pea, (*b*) maize, (*c*) marrow?

15. What conditions are necessary for the germination of seeds?

Chapter 7

1. Why do plants have roots?

2. With the help of diagrams, show the difference between a tap root and a fibrous root system.

3. What is meant by the words annual, biennial and perennial?

4. What is an adventitious root?

5. What is the difference between a celandine tuber and a potato tuber?

6. Write a brief essay on climbing plants.

7. Describe the structure of a crocus corm and explain what happens from the time that you plant the corm until the leaves are dead.

8. Why is couch grass a very difficult weed to destroy?

9. Compare and contrast the structure of the rhizomes of Solomon's seal and iris.

10. Of what use are the runners of a strawberry plant?

11. Explain how a blackberry stolon is formed. How could you recognize a stolon?

12. Briefly describe the difference between heart wood and sap wood.

13. With the help of a diagram, describe the structure of a twig.

14. Describe the structure of a horse chestnut bud.

15. How would you recognize the twigs of poplar, lime, sycamore, ash and beech?

16. What is a deciduous plant.

17. How could you tell the age of a twig and of a tree?

18. What is the difference between (*a*) simple and compound leaves, (*b*) net veined and parallel veined leaves?

19. Simply describe the structure of a leaf.

20. Describe the structure of a bulb. How does it differ from that of a corm.

APPENDIX A

APPARATUS AND MATERIALS
REQUIRED

The practical work in this book can be carried out with very little apparatus. Jam jars are extremely useful; and plain glass, screw-topped jars can be used for preserving specimens. Insect boxes can be bought or made (see Appendix B).

The following items are needed to study small specimens:

hand lenses ⎫
mounted needles ⎪
watch glasses ⎬ one between two
black and white tiles ⎭

1 doz. pipettes with rubber teats
1 microscope (or more) with two eye-pieces and 2 objectives
$\frac{1}{2}$ gr. microscope slides 3 in. × 1 in.
1 doz. well slides
$\frac{1}{2}$ oz. cover slips $\frac{1}{2}$ in. diam. medium thickness.

To study pond life it is necessary to have several aquaria. An electric aerator is useful but not necessary.

To preserve specimens the following items are required:

1 w.qt. formaldehyde 40 %.
4 doz. specimen tubes 2 in. × 1 in.
1 doz. specimen tubes 3 in. × 1 in.
1 doz. specimen tubes 4 in. × 1 in.
8 oz. chloroform.

Specimens

As far as possible children should find their own specimens. Names of dealers in specimens of livestock and microscope slides may be obtained from *The School Nature Study Union Journal*, which is issued in January, April, July and October. The annual subscription is five shillings and is due on 1 January. Subscriptions should be sent to:

THE HON. SUBSCRIPTION SECRETARY,
DR WINIFRED PAGE,
5 DARTMOUTH CHAMBERS,
THEOBALD'S ROAD, LONDON

APPENDIX B

SOME USEFUL HINTS

Siphoning

Place a jar of water above the aquarium. Take a piece of bent glass tubing with one arm reaching to the bottom of the jar. The other arm must be long enough to reach below the bottom of the jar on the outside. Suck the air out of the tube and the water will run freely. You can use this method to empty an aquarium.

The microscope

A microscope is an instrument used to make minute objects appear very much enlarged, so that these objects can be easily viewed by the eye. It consists of a combination of two lenses in a tube, one known as the OBJECTIVE and the other as the EYEPIECE. An ordinary standard school-type microscope, as shown in Fig. 139, with two eye-pieces and two objectives is useful for elementary work. The microscope must be kept in a special case to keep it free from dust.

How to use the microscope

(*a*) Place the slide on the platform.

(*b*) Move the mirror until the light is reflected through the object.

(*c*) Move the *coarse* adjustment slowly, until you can see the object clearly.

(*d*) *High-power objective.* After finding the object with the low-power objective, turn round the nose piece until the high-power objective is under the draw tube. Then bring the object into focus by carefully turning the *fine* adjustment.

N.B. Do not use the high-power objective unless the object is covered with a cover slip.

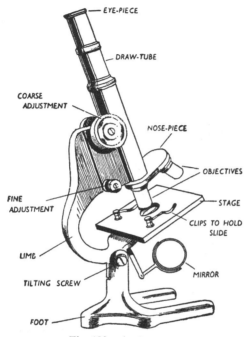

EYE-PIECE

DRAW-TUBE

COARSE ADJUSTMENT

NOSE-PIECE

OBJECTIVES

FINE ADJUSTMENT

STAGE

CLIPS TO HOLD SLIDE

LIMB

MIRROR

TILTING SCREW

FOOT

Fig. 139. *A microscope*

Microscope slides

Much time is necessary to acquire the technique of making permanent slides, and this is not a suitable type of work for the pupils who are not studying biology at an advanced level. It is, therefore, advisable to buy permanent slides. Temporary slides can be made in the following way:

1. Put your specimen on to the middle of a clean slide, and drop a spot of water on to it.

207

2. Put one edge of a clean coverslip on the edge of the drop of water, and lower the coverslip slowly, with the help of a mounted needle, being careful not to get an air bubble in the water.

Trout hatchery

If you wish to see the development of fish in your school, obtain the pamphlet on *Trout Hatchery in School*, which is published by the School Nature Study Union.

Insect boxes

Insect boxes are expensive to buy but you can make your own in the following way. Cut the front out of a cardboard box, leaving an edging of about one inch all the way round. Make the back of the cage to open, and make a fastener. Cut out a rectangle and over this stitch a piece of perforated zinc. When this is done, cover the front of the box with cellophane paper. Then paint the box. If the box is very strong you could use glass instead of cellophane paper.

Glass insect cages can be made by fastening four pieces of glass together with strong adhesive tape. Plasticene will secure this to a cardboard base, and the top may be made of perforated zinc.

Preserving specimens

This may be done in two ways:

(*a*) *Wet.* Place the dead specimen in a tube or in a bottle of appropriate size containing a two per cent. solution of formaldehyde.

(*b*) *Dry.* Small dead animals can be kept in insect boxes. These may be bought from dealers, or you can make wooden boxes with glass lids, or you can make them with match boxes or tins that have no lids. These can be covered with cellophane paper.

As soon as the animal is dead it should be spread out into the required position before it becomes stiff. You may either rest the specimen on cotton wool inside the box, or you can stick a pin through the thorax of the animal. If pins are used, you must put a layer of cork on the bottom of the box, into which the pin may be stuck. Place a little naphtha, or moth balls or a similar substance in the boxes to preserve the animals.

To kill animals

If you wish to preserve animals that are not dead, kill them first by putting them into a killing bottle. Special killing bottles can be bought from dealers or you can make your own in the following way. Put some cotton wool that has been soaked in chloroform in the bottom of a wide-necked bottle. Then place some dry cotton wool on top of this.

Specimens

In addition to the specimens mentioned in the book, the following may be found useful:

Reptiles. Nearly all English reptiles can be kept in school in a suitable vivarium (see Fig. 63).

Dead specimens may be pickled in five per cent. formaldehyde, and are very useful when living specimens are not available. Children could try to find sloughed skins of snakes.

Birds. Many useful specimens may be collected by the children and brought to school. For instance: (*a*) feathers of all kinds; (*b*) skeletons of birds whose flesh has been eaten; (*c*) skeletons of dead birds that have been buried in the ground; (*d*) teachers must give advice on the advisability of collecting birds' eggs and nests.

Mammals. A very wide range of specimens may be brought by the children. Dead specimens of small mammals or baby animals may be found useful if pickled in five per cent. formaldehyde. Skulls, hoofs and bones of all kinds will be collected by children. Do not keep living mammals unless you have suitable living accommodation for them.

Tree studies

Leaf tracings. Place a leaf under a piece of plain paper and hold it firmly. Scribble on the paper above the leaf using a pencil of the same colour as the leaf. Cut out the traced leaf and fasten it in your book.

Plaster casts. When you are studying trees, you can easily collect leaves, twigs, flowers and fruits, but you cannot cut pieces out of the bark, to complete your study. So try to make plaster casts of the bark. Take a piece of plasticene, and flatten it to the required size. Smear it with vaseline and press it on to the bark of the tree. Carefully remove it, and place it in a tin or dish. Then mix up plaster of Paris and put on top of the plasticene mould. Leave it to set and then remove the cast from the plasticene. Paint it the same colour as the bark, and you will have a cast that really looks like a piece of bark. Casts may be made of leaves, twigs, fruits, footprints, etc.

APPENDIX C

The following books should be included in the School science library, and used by the children to identify plants and animals that they find which are not mentioned in this book.

BATTEN, H. MORTIMER. *Our Garden Birds.*
BENTHAM AND HOOKER. *A British Flora.* 2 vols.
BOULENGER, E. G. *A Naturalist at the Zoo.*
BOULENGER, E. G. *Zoo Cavalcade.*
BOULENGER, E. G. *World Natural History.*
BOULENGER, E. G. *The Aquarium.*
BRIMBLE, L. J. F. *Intermediate Botany.* 3rd ed.
BRIMBLE, L. J. F. *Trees in Britain.*
BUCHSBAUM. *Animals without Backbones.*
COWARD, T. A. *Life of the Wayside and Woodland.*
COWARD, T. A. *Birds of the British Isles and their eggs.* 3 vols.
DAYLISH, E. TITCH. *Name this Bird.*
DUNCAN, E. MARTIN. *Cassell's Natural History.*
FURNEAUX, W. *Life in Ponds and Streams.*
GROOM, PERCY. *Elementary Botany.*
HELLYER, A. G. L. *Garden Pest Control.*
IMMS, A. D. *Insect Natural History.*
JENKINS, J. T. *Fishes of the British Isles.*
JOHNS, REV. C. A. *Flowers of the Field.*
JOY, NORMAN H. *British Beetles. Their Homes and Habits.*
KELMAN, J. H. *Butterflies and Moths.*
KELMAN, J. H. *Bees.*
LULHAM, R. *An Introduction to Zoology through Nature Study.*
MELLANBY, H. *Animal Life in Fresh Water.*
POCHIN, E. *How to recognize Trees of the Countryside.*
ROMER, A. S. *Man and the Vertebrates.*
SIDEBOTHAM, H. *Wild Animals.*

STEP, EDWARD. *British Insect Life.*

STEP, EDWARD. *Bees, Wasps, Ants and Allied Insects of the British Isles.*

STEP, EDWARD. *Wild Flowers Month by Month in their Natural Surroundings.* 2 vols.

STEP, EDWARD. *Wayside and Woodland Blossoms.* Series I, II and III.

THOMPSON, H. S. *How to collect and dry Flowering Plants.*

TINN, FRANK. *Eggs and Nests of British Birds.*

WATTS, W. M. *A School Flora.*

PUFFIN BOOKS:

   Animals of the Countryside.

   Trees in Britain.

   The Story of Plant Life.

   A Book of Insects.

   Butterflies in Britain.

'OBSERVER'S' BOOKS:

   British Wild Animals.

   British Butterflies.

   British Wild flowers.

   Trees and Shrubs of the British Isles.

'HOW TO RECOGNIZE' BOOKS:

   POCHIN, E. *British Wild Birds.*

   POCHIN, E. *British Birds, Eggs and Nests.*

'FOR THE POCKET' BOOKS—Oxford:

   SANDARS, EDMUND. *A Butterfly Book.*

   SANDARS, EDMUND. *A Beast Book.*

   SANDARS, EDMUND. *A Bird Book.*

POCKET GUIDE SERIES—Warne:

   Wild Flowers of the Wayside and Woodland. Compiled by T. N. SCOTT and W. J. STOKOE.

   Birds of the Wayside and Woodland. T. A. COWARD.

   Butterflies and Moths of the Wayside and Woodland. Compiled by W. J. STOKOE.

YOUNG FARMERS' CLUB BOOKLETS:

   1. *The Farm.* 2. *Rabbit Keeping.* 3. *Pig Keeping.* 4. *Poultry Keeping.* 5. *Bees.* 6. *Goat Keeping.* 9. *Garden and Farm Pests.* 10. *Cows and Milk.* 11. *Ducks, Geese and Turkeys.* 13. *Farm Horses.* 16. *Sheep Farming.*

MINISTRY OF AGRICULTURE
AND FISHERIES

A list will be sent, on application to the Ministry, of all the books and pamphlets that can be bought on the subjects of birds, keeping and rearing of animals, gardening, weeds, garden pests, etc.

INDEX

Printed in the United States
By Bookmasters